NATURAL DISASTER RESEARCH, PREDICTION AND MITIGATION

TSUNAMI HAZARD ASSESSMENT AT U.S. NUCLEAR POWER PLANTS

NATURAL DISASTER RESEARCH, PREDICTION AND MITIGATION

Additional books in this series can be found on Nova's website under the Series tab.

Additional E-books in this series can be found on Nova's website under the E-book tab.

NATURAL DISASTER RESEARCH, PREDICTION AND MITIGATION

TSUNAMI HAZARD ASSESSMENT AT U.S. NUCLEAR POWER PLANTS

Roger F. Goldman
EDITOR

Nova Science Publishers, Inc.
New York

For permission to use material from this book please contact us:
Telephone 631-231-7269; Fax 631-231-8175
Web Site: http://www.novapublishers.com

Additional color graphics may be available in the e-book version of this book.

Library of Congress Cataloging-in-Publication Data

Tsunami hazard assessment at U.S. nuclear power plants / editor, Roger F. Goldman.
p. cm.
Includes bibliographical references and index.
ISBN 978-1-61470-865-0 (hardcover)
1. Nuclear power plants--Risk assessment--United States. 2. Tsunami damage--Risk assessment--United States. I. Goldman, Roger F.
TK9152.16.T77 2011
363.17'992--dc23

2011028026

Published by Nova Science Publishers, Inc. † New York

CONTENTS

PREFACE

This book examines the tsunami phenomenon with the focus on its relevance for hazard assessment at nuclear power plant sites. Three tsunamigenic mechanisms (earthquakes, landslides, and volcanoes) are considered. Also discussed are offshore and onshore modeling of tsunami waves and their effects on nuclear power plant structures, systems, and components. Identification of the repository of historic tsunami data and ways to approach site safety assessment for tsunamis by a Nuclear Regulatory Commission is also reviewed.

Chapter 1- The U.S. Nuclear Regulatory Commission (NRC) Office of New Reactors (NRO) is responsible for the licensing and regulatory oversight of civilian nuclear power reactors and research reactors in the United States. Commission Paper SECY-06-0019 (Reyes 2006) described how recent developments within the United States and abroad have contributed to increased interest in licensing and construction of new reactors. Nuclear energy is also encouraged by the 2005 Energy Act, which contains a provision for U.S. Government standby support for any delays due to NRC reviews. In the Staff Requirements Memorandum, dated May 10, 2005, from the April 6, 2005, Commission Briefing on Status of New Site and Reactor Licensing, the Commission requested that the staff update licensing-guidance documents, including the Standard Review Plan (SRP), NUREG-0800 (NRC 1996) to support new reactor licensing. The Pacific Northwest National Laboratory (PNNL) assisted the NRC in updating the SRP. The updated SRP was released by the NRC in March, 2007 (NRC 2007).

Chapter 2- A 9.0 magnitude earthquake struck off Japan's northeast coast near Honshu in the afternoon on Friday, March 11, 2011 (12:46 a.m. eastern time in the United States). The earthquake triggered a tsunami that has caused widespread devastation to parts of the coastal regions in Japan closest to the earthquake. The tsunami traveled across the Pacific Ocean, and the National Oceanic and Atmospheric Administration (NOAA) tsunami warning centers in Hawaii and Alaska issued tsunami warnings for coastal areas of Hawaii, Guam, the Commonwealth of the Northern Marianas, American Samoa, Alaska, and California. Although the tsunami caused widespread damage along the northeast coast of Japan, tsunami warnings issued from the tsunami warning centers gave the above U.S. Pacific territories, Hawaii, and the U.S. West Coast adequate warning to prepare for incoming waves.

In: Tsunami Hazard Assessment at U.S. Nuclear... ISBN: 978-1-61470-865-0
Editors: Roger F. Goldman

Chapter 1

TSUNAMI HAZARD ASSESSMENT AT NUCLEAR POWER PLANT SITES IN THE UNITED STATES OF AMERICA

United States Nuclear Regulatory Commission

ABSTRACT

We describe the tsunami phenomenon with the focus on its relevance for hazard assessment at nuclear power plant sites. Chapter 1 includes an overview of tsunamis and mechanisms that generate tsunamis. Three tsunamigenic mechanisms—earthquakes, landslides, and volcanoes—are considered relevant for hazard assessment at nuclear power plant sites. We summarize historical tsunami occurrences, including descriptions of source mechanisms and damages caused by these events. Historical landslides and potential landslide areas in earth's oceans are described. We describe the hierarchical-review approach to tsunami-hazard assessment at nuclear power plant sites in Chapter 2. The hierarchical-review approach consists of a series of stepwise, progressively more-refined analyses to evaluate the hazard resulting from a phenomena at a nuclear power plant site. We recommend that the hierarchical- review approach employ a screening analysis to determine if a site is subject to tsunami hazard based on the presence of a tsunamigenic source and the location and elevation of the site. The screening analysis is expected to ensure that analysis and review resources are not wasted at sites with little potential of exposure to tsunamis. Chapter 3 describes the effects tsunami waves may have at a nuclear power plant site. These effects result in hazards that may directly affect the safety of a plant's structures, systems, and components. Structures, systems, and components important to the safety of a plant should be adequately designed and, if required, protected from these hazards. Chapter 4 describes data required for a detailed tsunami-hazard assessment and sources of these data. We recommend using existing resources and previously completed tsunami-hazard assessments, if available and appropriate. Detailed tsunami-hazard assessment at a nuclear power plant site should be based on the probable maximum tsunami. Chapter 5 defines the probable maximum tsunami, its determination at a site, and subsequent hazard assessment. We point out that a tsunamigenic source that produces probable maximum tsunami hazards at a site may not be determined a priori. It may be necessary to evaluate several candidate sources and the tsunamis generated from them under the most favorable tsunamigenic source and ambient conditions. The set of hazards obtained from all such scenario tsunamis should be considered to determine design bases of the plant structures, systems, and components. Chapter 6 describes international practices by Japan and the International Atomic Energy Agency, which are reviewed for completeness. The appendix provides a stepwise guide to site-independent analyses for tsunami-hazard assessment.

FOREWORD

The Indian Ocean tsunami of December 2004 led the NRC to take the initiative to examine its criteria for nuclear plant siting evaluation against tsunami hazards. Site evaluation against tsunami hazards, as incorporated into the Standard Review Pan, was determined to be comprehensive; however, it needed to be updated to incorporate current understanding of tsunami sources, modeling near shore and on-shore wave surge, draw-down, erosion and other associated effects. In 2005, the National Tsunami Hazard Reduction Program was initiated by the President, and the current study was conducted by the Pacific Northwest National Laboratory in collaboration with the Pacific Marine Environmental Laboratory (PMEL) of the National Oceanic and Atmospheric Administration (NOAA) to ensure that NRC's guidance on site evaluation against tsunami hazards are consistent with the National Program.

The focus of this study has been to examine tsunami hazards at nuclear power plant sites, to review offshore and onshore modeling of tsunami waves, to describe the effects of tsunami waves on nuclear power plant structures, systems, and components, to develop potential approaches for screening sites for tsunami effects, to identify the repository of historic tsunami data, and to examine the ways to approach site safety assessment for tsunami for an NRC reviewer.

NRC's Office of Nuclear Regulatory Research (RES) has initiated a comprehensive research program on source characterization, modeling, tsunami effects and probabilistic hazard framework, as appropriate. The RES program is intended to fill in the information gaps that exist on tsunami sources for the Atlantic and the Gulf coast areas of the continental United States, and limited characterization of submarine deposits that can slide and cause landslide induced tsunami. Results from the RES program, combined with technical tsunami modeling software from NOAA PMEL will enable comprehensive tsunami hazard assessment at future nuclear power plant sites following the site safety assessment approach described here.

James E. Lyons, Director
Division of Site & Environmental Reviews
Office of New Reactors

Executive Summary

The U.S. Nuclear Regulatory Commission (NRC) Office of New Reactors (NRO) is responsible for the licensing and regulatory oversight of civilian nuclear power reactors and research reactors in the United States. Commission Paper SECY-06-0019 (Reyes 2006) described how recent developments within the United States and abroad have contributed to increased interest in licensing and construction of new reactors. Nuclear energy is also encouraged by the 2005 Energy Act, which contains a provision for U.S. Government standby support for any delays due to NRC reviews. In the Staff Requirements Memorandum, dated May 10, 2005, from the April 6, 2005, Commission Briefing on Status of New Site and Reactor Licensing, the Commission requested that the staff update licensing-guidance documents, including the Standard Review Plan (SRP), NUREG-0800 (NRC 1996) to support new reactor licensing. The Pacific Northwest National Laboratory (PNNL) assisted the NRC in updating the SRP. The updated SRP was released by the NRC in March, 2007 (NRC 2007).

Licensing of nuclear power plants is regulated by Title 10 of Code of Federal Regulation (CFR) Parts 50, 52, and 100. Appendix A to 10 CFR Part 50, General Design Criteria for Nuclear Power Plants, contains criterion 2 (GDC 2), which requires that structures, systems, and components important to safety be designed to withstand the effects of natural phenomena, including tsunamis. A series of regulatory guides (RG) provide guidance to licensees and applicants on implementing specific parts of the NRC's regulations, techniques used by the NRC staff in evaluating specific problems or postulated accidents, and data needed by the staff in its review of applications for permits or licenses. An update to RG 1.59 (NRC 1977) was expected to include guidance for assessment of tsunamis as a flooding hazard, but was never completed. An update to RG 1.59 is currently underway.

In the wake of the December 26, 2004, Sumatra earthquake and its accompanying tsunami that resulted in widespread loss of life and property in the Indian Ocean region, hazards posed by tsunamis have emerged as some of the most severe caused by natural phenomena. One operating nuclear power reactor was shut down during this tsunami, and, therefore, international nuclear power plant operators and reviewers felt the need to review the approach towards tsunami-hazard assessment for existing and proposed sites. The President of the United States launched an initiative to improve domestic tsunami-warning systems, and its first installment was signed into law on May 11, 2005 (STC 2005). In April, 2006, the NRC requested technical assistance from Pacific Northwest National Laboratory in the development of new guidance documents for tsunami-related hazards at nuclear power plant sites, consistent with the President's initiative.

Section 2.4.6 of the SRP (NRC 2007) describes review procedures and acceptance criteria for probable maximum tsunami (PMT) hazards. However, technical guidance related to methods and data required for tsunami-hazard assessment is not the focus of the SRP. The National Oceanic and Atmospheric Administration (NOAA) has the responsibility to develop standards of accuracy for tsunami-simulation models and to conduct research to support the National Tsunami Hazard Mitigation Program. NOAA's responsibilities were reaffirmed on December 20, 2006, when the President signed the Tsunami Warning and Education Act. NOAA's Pacific Marine Environmental Laboratory (PMEL) operates the NOAA Center for Tsunami Research (NCTR). NCTR's objectives include the development and implementation of improved models to increase the speed and accuracy of operational forecasts and warnings, and development of improved methods to predict the impacts of tsunamis on coastal communities and infrastructure. To leverage the extensive technical expertise and regulatory authority that NOAA has and PNNL's experience with reviewing the first three Early Site Permit applications, PNNL and NRC agreed to develop guidance for tsunami-hazard assessment using a set of two documents—this report, prepared by PNNL, that focuses on the review guidance relevant for NRC staff and a PMEL-prepared NOAA Technical Memorandum titled "Scientific and Technical Issues in Tsunami Hazard Assessment of Nuclear Power Plant Sites" (Gonzalez 2007) to serve as the technical basis. As tsunami-simulation models are developed, standardized, and released by PMEL for use by the tsunami community, the NOAA Technical Memorandum will be updated to include the state- of-the-art in tsunami-hazard assessment.

A hierarchical approach is recommended for review of tsunami hazards at nuclear power plant sites. An initial screening should be performed to establish if the site is free from tsunamis. A general rule based on horizontal distance (D), longitudinal distance along a stream from an estuary (L), and elevation (Z) of structures, systems, and components (SSC) important to safety could be used to reasonably demonstrate that a plant may be safe from any tsunami hazards. Because tsunami runup and drawdown, and the associated hazards, are highly site-specific, a priori numerical limits on D, L, and Z cannot be established. Site-specific analysis of data combined with sound engineering judgment may be used to justify that detailed tsunami hazard assessment at the site may not be required. It is expected that the majority of the sites located far from the shoreline may be screened out. If the initial screening indicates that tsunami hazards are of concern, a detailed assessment of probable maximum tsunami (PMT) hazards should be performed. Generally, it may not be possible to a priori determine the tsunamigenic source and associated parameters that will result in PMT hazards at a site. Therefore, a set of candidate tsunamis generated by all possible tsunamigenic sources should be simulated to determine the worst hazard at the site. Because tsunami-wave characteristics are highly dependent on near-shore bathymetry and geometry of inlets, bays, and coves, the hazards are expected to be highly spatially variable in the near-shore area. Maps of worst-case PMT-wave characteristics and hazard metrics should be compiled from the scenario tsunami simulations and used to determine design bases and any required protection for affected SSC important to safety.

At this time, efforts at NOAA PMEL are underway to develop a community tsunami- modeling system. The standards, criteria, and procedures for evaluation of tsunami numerical models have been published by NOAA in

Technical Memorandum OAR PMEL-135. The community tsunami-modeling system would greatly aid the estimation of PMT hazards.

ACKNOWLEDGMENTS

Early and enthusiastic support for the initiative for this study was provided by Dr. Brian Sheron, Director of RES, Mr. Michael Mayfield, Director, Division of Engineering, NRO, Dr. Nilesh Chokshi, Deputy Director, Division of Site and Environmental Reviews, NRO and Dr. Eddie Bernard, Director, PMEL, NOAA.

The author wishes to sincerely acknowledge technical discussions, guidance, and suggestions provided by Mr. Lance Vail, PNNL. Comments provided by Mr. Goutam Bagchi, Dr. Hosung Ahn, Dr. Christopher Cook, Dr. Ann Kammerer, and Mr. Mark McBride, all of NRC, helped improve this report. Dr. Frank Gonzalez and Dr. Vasily Titov, both of NOAA PMEL, graciously provided comments on technical aspects of tsunamis.

Mr. Stuart Saslow, Ms. Janie Vickerman, Ms. Julie Hughes, Mr. Kevin Ghirardo, and Ms. Liz Davis, all of PNNL, provided excellent project support. Editorial review of this report was performed by Ms. Sheila Bennett and Ms. Angie Aguilar, PNNL.

Financial support from the Division of New Reactor Licensing, Office of Nuclear Reactor Regulation, and later, the Office of New Reactors, NRO, of the NRC is gratefully acknowledged. Preparation of this report would have been impossible without the continued guidance and support of Mr. Goutam Bagchi, the NRC Technical Monitor.

1. TSUNAMI AND OTHER TSUNAMI-LIKE WAVES

1.1. Introduction

The word tsunami is a Japanese word that literally means "harbor Ctsul wave (`nami')." A tsunami is a series of water waves that propagate from the point of generation (the location of the tsunamigenic source) toward the shore. Typically, the term tsunami refers to an oceanic tsunami caused by the initiation of the tsunami waves due to the vertical displacement of the water column by some submarine tsunamigenic source. However, tsunami or tsunami-like waves can also be generated in inland water bodies by sources that have appropriate tsunamigenic characteristics.

Tsunamis can be severely destructive to infrastructure, human life, and the economy located near the coast. These waves can travel great distances in the form of gravity waves with little loss of energy. The waves increase in amplitude as they reach shallow water near the shore. The waves may inundate large areas onshore depending on local bathymetry and topography. The hydrostatic and hydrodynamic forces associated with the waves can damage structures. The structures on land may also be impacted by water-borne debris and projectiles.

Tsunamis are of great interest to nuclear power plants located near a shoreline that may be affected by tsunamis. Tsunami hazards that may affect safety should be considered in the siting of the plant. The design of the SSC important to safety should also consider tsunami hazards to ensure that the threat posed to the plant and, subsequently, to public health and safety are adequately mitigated in the design bases.

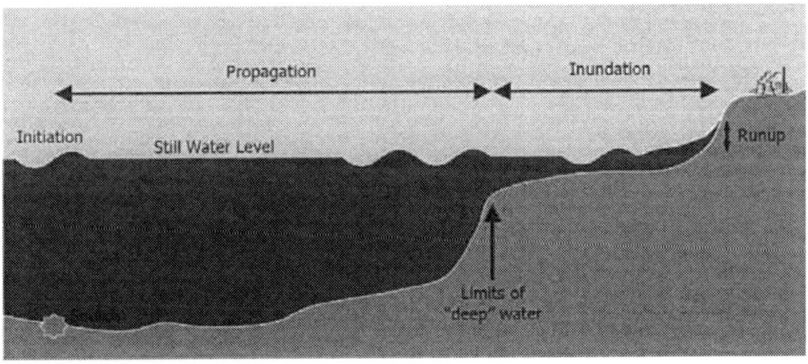

Figure 1-1. A far-field oceanic tsunami.

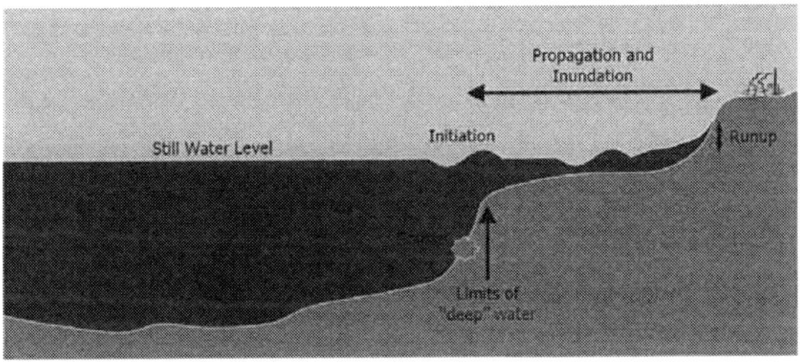

Figure 1-2. A near-field oceanic tsunami.

There are three distinct "phases" of a tsunami, as illustrated in Figure 1-1. A far-field tsunami is one for which the source is located more than 625 mi (1000 km) from the area of interest.

Little energy is lost during propagation of the tsunami waves, and the speed at which these waves travel can be estimated based on the theory of shallow-water waves. The propagation phase of the tsunami waves can be approximated well with linear theory. The final phase of the tsunami is the inundation phase, where the waves enter shallower waters near the shore, shorten in wavelength, and increase in amplitude. Nonlinear effects become significant during the inundation phase and cannot be neglected. The amplification of the wave depends on the local, near-shore bathymetry. The geometry of the shoreline— combined with bathymetry—can also lead to reflection, refraction, trapping of waves, and other interactions that may further modify the characteristics of the tsunami waves. The runup is defined as the maximum ground elevation that he tsunami waves reach above a datum.

Figure 1-2 shows a near-field oceanic tsunami. The source for a near-field tsunami is generally less than 625 mi (1000 km) from the area of interest. Due to the proximity of the source to the shore, the waves arrive at the area of interest quickly, which may limit the time available for evacuation or protective actions.

Figure 1-3 shows a tsunami generated in an inland water body, such as a lake or a man-made reservoir, initiated by a seismic source beneath the water body. Other causes, like hillslope failures and subaerial landslides, may also be appropriate as a source mechanism for tsunami-like waves generated in inland water bodies. Hillslope failure is the term generally used to describe a slope failure caused by increased pore-water pressure that results in loss the of sheer strength in the soil and may lead to sliding along weakened layers. Subaerial landslide is the term generally used to describe a landslide that starts above the surface of the water body and impacts the water body, generating a tsunami-like wave.

1.2. Definition

The National Oceanographic and Atmospheric Administration defines tsunamis as "...a series of very long waves generated by any rapid, large- scale disturbance of the sea."

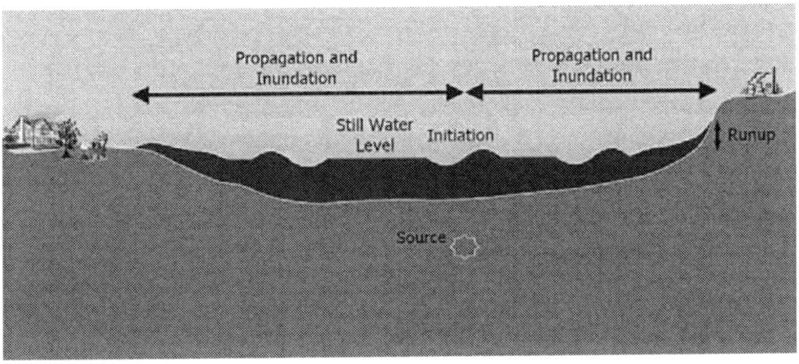

Figure 1-3. A tsunami in an inland water body.

The International Tsunami Information Center of the Intergovernmental Oceanographic Commission—a part of the United Nations Educational, Scientific, and Cultural Organization—defines tsunamis as "...a series of traveling waves of extremely long length and period, usually generated by disturbances associated with earthquakes occurring below on near the ocean floor."

For the purposes of this report, we adopted the following definition:

> A tsunami is a series of water waves generated by a rapid, large-scale disturbance of a water body due to seismic, landslide, or volcanic tsunamigenic sources.

Note that this definition is not limited to oceanic tsunamis. Landslide sources may include submarine and subaerial slides and ice falls. Volcanic sources include the effects of pyroclastic flows and caldera collapse. The effects of a caldera collapse may be similar to a submarine or a subaerial landslide, depending on the location and the characteristics of the event.

1.3. Mechanisms

Tsunamis are generated by rapid, large-scale disturbance of a body of water. Therefore, only geophysical events that release a large amount of energy in a very short time into a water body generate tsunamis. The most frequent cause of tsunamis is an earthquake. Less frequently, tsunamis are generated by submarine and subaerial landslides, by pyroclastic flows and caldera collapses during volcanic eruptions, meteorite impacts, and by ice falls.

1.3.1. Earthquakes

The most frequent source of tsunami generation is a submarine earthquake. Earthquakes primarily generate tsunamis through vertical displacement of the floor of the water body that results in a simultaneous (often assumed identical) displacement of the overlying water column. Because of the tsunami-generation sequence associated with earthquakes, dip-slip earthquakes are more efficient at generating tsunamis than strike-slip earthquakes. However, Tanioka and Satake (1996) show that it is possible for a strike-slip fault to generate major tsunamis under certain conditions where horizontal displacement of a steep slope leads to a significant vertical displacement of the water column.

To generate a major tsunami, a substantial amount of slip and a large rupture area is required. Consequently, only large earthquakes with magnitudes greater than 6.5 generate observable tsunamis. The controlling source parameter that determines tsunami severity is the seismic moment, M_0, defined as

$$M_0 = \mu \overline{D} A$$

where μ is the shear modulus or rigidity, $\overline{D}$ is the average slip, and A is the rupture area. Moment magnitude of the earthquake, M_w, is computed through the empirical relationship

$$M_w = \frac{2}{3}\left[\log(M_0) - 9.05\right]$$

The generation of the tsunami from an earthquake event is carried out in three steps:

1. fault rupture modeling
2. calculation of coseismic displacements
3. calculation of initial wave field.

Fault-Rupture Modeling

The most commonly used rupture model is based on elastic dislocation. Elastic dislocations under the assumption of uniform slip are called Volterra dislocations. For a given earthquake magnitude M_w, rupture area A, and shear modulus μ, it is possible to estimate an equivalent uniform slip. However,

Geist (2005) explains that literature demonstrates that the assumption of uniform slip implies that the deformation is concentrated at the edges of the rupture. Relaxation of the uniform-slip assumption to allow for variable slip in the dip direction for a dip-slip fault results in concentration of the deformation near the center of the rupture zone and substantially higher vertical displacements (Geist and Dmowska 1999), leading to a greater initial tsunami wave height.

The less frequently used rupture model is based on crack theory (Geist and Dmowska 1999), in which a uniform static stress drop, rather than a uniform slip, is specified. The seismic moment and the uniform stress drop are related by (Lay and Wallace 1995)

$$M_0 = \frac{3}{8}\pi\,\Delta\sigma\,W^2\,L$$

where $\Delta\sigma$ is the uniform static stress drop, W is the width of the rupture, and L is the length of the rupture.

Analysis of seismic-waveform data and other thebretical and numerical studies of rupture dynamics have shown that slip distribution is strongly heterogeneous, such that the slip distribution can rarely be considered uniform and conforms to the crack theory only in certain cases (Yomogida 1988). For far-field tsunamis, the effects of slip distribution are attenuated by the time the tsunami reaches the site. However, for near-field tsunamis, care should be taken to use an appropriate rupture model because slip heterogeneity has a significant effect on tsunami- wave heights. For an arbitrary slip distribution, discretized cells of uniform dip-slip and the point source expressions of Okada (1985) can be used. Geist and Dmowska (1999) show the effects of slip heterogeneity modeled using discretized cells on the local tsunami wave field.

Calculation of Coseismic Displacements

Coseismic displacement of the floor of the water body can be estimated using analytical expression of Okada (1985) for homogeneous earth structure and a rectangular planar fault for cases where the rupture is represented by a Volterra elastic dislocation. Analytic expressions for coseismic displacements have been developed for crack models of rupture using Chebyshev polynomials (Dmowska and Kostrov 1973; Rudnicki and Wu 1995). In general, a heterogeneous slip field can be discretized into cells of planar, uniform slip using the point source expressions of Okada (1985). These techniques apply to planar faults in a homogeneous elastic medium only.

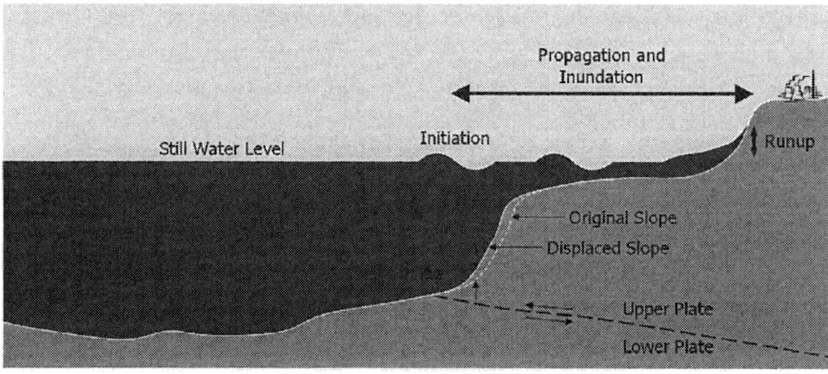

Figure 1-4. Vertical component of displacement due to a shallow dipping thrust.

For non-planar faults, Jeyakumaran et al. (1992) and Jeyakumaran and Keer (1994) developed analytical expressions that use triangular dislocations and curved slip zones. For non-planar faults in heterogeneous medium, numerical techniques, like finite elements, or boundary elements may also be used (Yoshioka et al. 1989; Zhao el al. 2004).

Calculation of Initial Wave Field

The vertical component of the coseismic displacement of the floor of the water body dominates tsunami generation. Therefore, dip-slip earthquakes are more efficient in tsunami generation. Tanioka and Satake (1996) also show, however, that tsunamis can be generated under certain conditions by earthquakes in which horizontal displacement is large relative to the vertical displacement. Figure 1-4 shows a situation where the displacement occurs on a shallow dipping thrust. The horizontal movement of the upper plate to the left of its original position results in a significant vertical component, generating a tsunami wave. Tanioka and Satake (1996) report this situation for the June 2, 1994, Java, Indonesia, earthquake of M_y, 7.8, which generated a tsunami with a maximum runup of 37 ft (11.3 m).

If a strike slip fault results in horizontal displacement of a steep slope, a tsunami may also be generated by the vertical component of the displacement. Tanioka and Satake (1996) proposed this mechanism for the November 14, 1994, Mindoro earthquake of M_y, 7.1 that generated a tsunami with a maximum runup of 24 ft (7.3 m) in the Phillipines.

The vertical component of the displacement due only to the horizontal movement of the slope is expressed as (Tanioka and Satake 1996)

$$u_h = u_x \frac{\partial H}{\partial x} + u_y \frac{\partial H}{\partial y}$$

where x and y denote horizontal dimensions, u_z and u_y are the horizontal displacements, H is the water depth measured positive downward, and u_h is the vertical component of the displacement measured positive upward. The total vertical displacement of the floor of the water body for a fault motion that also has a vertical displacement, u_z is therefore $u_z + u_h$. Depending on the characteristics of the fault and slope displacement, u_h may contribute significantly to the total vertical displacement of the floor of the water body. Tanioka and Satake (1996) suggest that this phenomenon may explain some of the tsunamis generated by the so-called tsunami earthquakes that are relatively small in magnitude yet generate major tsunamis.

The time during which the rupture takes place and results in displacement of the bottom of the water body can be considered instantaneous relative to the propagation speed of tsunamis. Therefore, the initial tsunami wavefield mimics the vertical- displacement field of the bottom of the water body. Wavelength components of the vertical- displacement field that are less than approximately three times the water depth are attenuated through the water column, but this phenomenon is mostly a concern in shallow and surface faulting. The initial displacement of the water surface can be assumed identical to that of the bottom of the water body under the assumption that water is an incompressible fluid.

Source Parameters

Earthquake source parameters for tsunami generation are related to the characteristics of the dislocation. These parameters fall into two general categories: those that scale with the magnitude of the earthquake and those that relate to material properties of the rupture zone.

Magnitude distribution

The frequency distribution of the magnitude of earthquakes is given by the Gutenberg-Richter (G-R) power law (Kagan 2002):

$$\log[N(M_w)] = a - bM_w$$

where $N(M_w)$ is the number of earthquakes with magnitudes greater than M_w and a and b are parameters of the power law function. The slope of the distribution (b) has been fairly well established in literature, but there is considerable debate on how to specify the tails of the distribution. The distribution is truncated on the high end because of the requirements of the preservation-of-energy principle. Kagan (2002) described four common forms of the size distribution for large-magnitude earthquakes. There are essentially three approaches to define the shape of the G-R distribution for large magnitudes: (1) the characteristic model, which assumes that the largest earthquake occurs at approximately the same location and at approximately the same magnitude; the G-R distribution is specified up to the magnitude of the largest aftershock of the characteristic earthquake (Wesnousky 1994); (2) a regionally modified G-R relationship; a continuous distribution where the right tail of the distribution falls off rapidly at a rate greater than the G-R slope, b; and (3) a globally modified G-R relationship that is based only on tectonic boundary type (Bird and Kagan 2004). Some studies (Kagan and Jackson 1991, 1995; Rong et al. 2003; Okal et al. 2006) indicate that the characteristic model may not be valid. There is some concern that regionally modified G-R relationships may suffer because of a lack of sufficient earthquake data to reliably estimate their parameters. Because of this, Bird and Kagan (2004) proposed a global G-R relationship based only on tectonic boundary type. Also, for subduction zones, the corner magnitude[a] is very high $(9.58^{+0.48}_{-0.46})$ Thus, unless proven otherwise, it should be assumed that earthquakes with moment magnitude greater than 9 can occur. Bird and Kagan (2004) provide the corner magnitude for oceanic convergent boundaries as $8.04^{+0.52}_{-0.22}$.

Fault dimensions

Fault dimensions consist of the rupture length, the rupture width, and the slip amount (more accurately, the distribution of the slip amount). Although the dimensions of the rupture scale with seismic moment, past a certain magnitude, the width of the rupture saturates. The saturation width depends on the frictional stability of the rupture zone, which varies with depth (Scholz 1990). Rupture length and slip may continue to increase after the saturation of the rupture width with a corresponding increase in seismic moment. General scaling relationships for rupture dimension are available (Geller 1976; Kanamori and Anderson 1975; Wyss 1979).

---

[a] The slope of the G-R relationship increases for earthquakes of magnitudes greater than the corner magnitude, resulting in a rapid falloff of the distribution.

Dip

Dip is the angle between the fault plane and the horizontal plane. It is generally determined from geophysical studies or from analysis of past events.

Strike

Strike is the geographic orientation usually given as a compass direction of the fault plane. It is also determined from geophysical studies or from analysis of past events.

Slip vector

Slip vector refers to the direction and amount of the slip during the earthquake. Relative motion between the plates at a convergent boundary almost always has an oblique component that results in a compression component as well as a shear or transcurrent component of strain (Soofi and King 2002). The orientation of this relative motion can attain any value between 0° and 90°, called the angle of obliquity. Generally, oblique convergence results in strike-slip and dip-slip components, thereby partitioning the slip into these two directions. The amount of slip partitioning controls the orientation of the slip vector along the interplate thrust. In cases of a fully slip-partitioned subduction zone, the earthquake has a pure thrust mechanism that has the maximum tsunamigenic efficiency.

Slip distribution

As discussed above, the amount of slip along the rupture can be highly variable. The slip heterogeneity can have a significant effect on near-field tsunami amplitudes, but its effects are significantly dampened in the far field. Slip-distribution models that account for heterogeneity are generally based on self-affine properties of rupture dynamics (Andrews 1980; Mai and Beroza 2002). The average slip for these models is estimated from the overall seismic moment, and the falloff of the seismic wavenumber spectrum is estimated from far-field displacement spectrum observed on seismograms. Andrews (1980) showed that a slip- distribution spectrum that decays as k^{-2} in the wave number (k) domain is consistent with the widely observed ω^{-2} decay (Aki 1967) in the frequency (co) domain. Herrero and Bernard (1994) proposed the "k-square" model, in which the slip- distribution spectrum decays as k^{-2} beyond a corner wavenumber, k_c. The corner wavenumber was related to fault length. These slip-distribution models can be used to generate a suite of scenario patterns of slip distribution, which has been used in ground-motion studies (Berge et al. 1998; Somerville et al. 1999) and may also be used in tsunami source models.

Mai and Beroza (2002) point out that models of slip distribution based on self-affinity are essentially fractal and, therefore, contain no characteristic length scale to describe the size of asperities. Somerville et al. (1999) found that the number of asperities and their size increase with seismic moment. They suggested that it may be possible to constrain the simulated slip-distribution patterns from a "k-square" model using the correlation of the size and the number of asperities with the seismic moment.

There is also an indication that slip may exhibit larger fluctuations than that predicted by the Andrews (1980) model. Lavallée et al. (2006) suggests that random variables from the Lévy distribution may be suitable for generating scenarios under these circumstances.

Shear modulus

Shear modulus is the constant of proportionality in Hooke's Law that relates shear stress to shear strain. It is also described as the initial linear slope of the stress-strain curve for shear. There is a large variation in the value of the shear modulus due to the large variety of types of rocks that are present in subduction zones. Shear modulus can be measured (Saffer et al. 2001) or deduced from earthquake source-time functions (Bilek and Lay 1999, 2000).

The variation of shear modulus with depth was shown to have a large effect on tsunamigenic potential of earthquakes: Okal (1988) showed, using a theoretical study, that a one tenth seismic moment located in low-rigidity sedimentary rocks resulted in an order-of-magnitude increase in the initial tsunami amplitude. This phenomenon may also explain tsunami earthquakes that are relatively small in magnitude yet generate large tsunamis. If these earthquakes are located at a shallower depth near the trench, and the subduction zone consists of low-rigidity material, four conditions favorable to a more efficient and greater tsunami generation are possible:

1. release of seismic moment in low-rigidity rocks increases slip compared to the same moment release in a higher-rigidity material
2. shallow rupture initiation increases the possibility of the rupture of the floor of the water body, which also tends to increase the slip
3. shallow focal depth increases the coseismic displacement
4. greater water depth near oceanic trenches results in greater amplification of the tsunami waves during shoaling.

1.3.2. Landslides

There are two broad categories of landslides: (1) submarine or subaqueous landslides that are initiated and progress beneath the surface of the water body and (2) subaerial landslides that are initiated above the water and impact the water body during their progression or fall into the water body.

The movement of a large mass of the slide or the impact of the fall displaces the water in the direction of the movement and can lead to generation of a tsunami wave on the surface of the water body. Once the initial wavefield is generated, it propagates outward from the source region. A review of submarine landslides is given by Hampton et al. (1996).

Source types

Landslides occur in several ways, depending on the geologic composition of the slope, steepness of the slope, triggering mechanism, and pore-water pressure. There are five classes of slope movement (Varnes 1978): (1) falls, (2) topples, (3) translational and rotational slides, (4) lateral spreads, and (5) flows. Depending on the location, material properties, and properties of the trigger, a combination of these movements may occur. These combined movements are called complex slope movements.

The initial tsunami-wave generation is affected by the type and the time history of the slope movement. Therefore, it is important to identify these parameters of the landslide in the area of interest. In a given area, several types of landslide events may occur that are capable of generating tsunamis. For example, in Alaska, destructive local tsunamis have occurred due to subaerial landslides (e.g., Lituya Bay in 1958) and as a result of submarine landslides in Valdez Arm of the Prince William Sound that were triggered by the 1964 Great Alaska earthquake. In southern California, tsunamis have been generated in the geologic past by submarine mud flows in Santa Barbara Channel and by debris avalanches in Palos Verdes.

Subaerial landslides have also occurred in inland lakes (a tsunami-like wave in Spirit Lake caused by debris flow after eruption and collapse of Mount St. Helens dome in 1980) (Waitt and Pierson 1994) and man-made water-storage reservoirs (a tsunami-like wave in Vaiont Reservoir in Italy caused by a massive hillslope failure and resulting landslide into the reservoir) (Kiersch 1964; Hendron and Patten 1985).

Submarine landslides

Several mechanisms can trigger a submarine landslide. The most common of these is an earthquake, such as the 1929 Grand Banks (Fine et al. 2005),

1946 Aleutian (Lopez and Okal 2006), 1964 Valdez (Lee et al. 2003), and the 1998 Papua, New Guinea, (Satake and Tanioka 2003) tsuna mis. Often, landslides triggered by an earthquake, can occur very shortly after the earthquake such that the generated tsunami is affected by both source mechanisms (Johnson and Satake 1997; Satake and Tanioka 2003; Lopez and Okal 2006). Many of the events in the National Geophysical Data Center tsunami catalog that are attributed to landslides may have such a composite source: In other instances, the slope failure may occur several hours after the triggering earthquake (Seed et al. 1975).

Tsunami generation mechanism

Tsunami generation from landslides can, in principle, be modeled similar to that from an earthquake. The physics of the slide are used to estimate the displacement of the overlying water column to determine the initial tsunami wavefield. A *series* of papers by Jiang and LeBlond (1992, 1993, 1994) describe the physics of submarine mudslide and the waves it generated. Heinrich et al. (2001) used granular material to simulate slide dynamics.

Onset of shear failure of a slope under normal stress can be described by the Mohr-Coulomb failure criterion:

$$\tau = c' + (\sigma - u) \tan\phi$$

where τ is the shear stress, c' is the cohesion of the slope material, u is the normal stress, σ is the pore pressure, and φ is the angle of friction (repose angle). The analysis of mass movement after failure can be analyzed using models proposed in literature (e.g., Imran et al. 2001).

Landslide tsunamigenic sources have two properties that are different from the earthquake sources: (1) landslide-generated tsunamis have a very strong directivity in the direction of mass movement, and (2) the source cannot be considered to generate an instantaneous displacement of the still-water level due to the time it takes for the slide to evolve, during which the characteristics of the surface waves are affected. Because of the time-dependent nature of source evolution, a completely coupled model (e.g., Jiang and LeBlond 1994) or a kinematic model (Lynett and Liu 2002) is generally preferred for modeling of landslide-generated tsunamis.

Initial wave characteristics

The outgoing wave from the landslide source propagates in the direction of the slide with its amplitude affected by the terminal velocity of the movement (Trifunac et al. 2002, 2003). The characteristics of the backgoing wave depends on the acceleration of the slide and appears as a depression wave approaching the shore. Two parameters of the slide primarily affect tsunami generation: the volume and time history of the slide.

Volume distribution

The frequency distribution of submarine landslide volume follows a power law similar to that for terrestrial landslides (Issler et al. 2005; ten Brink et al. 2006). The volume distribution is particularly important in estimating the large landslides; however, care should be exercised in the estimation because a large slide may be a composite of multiple smaller failures (Lee et al. 2004). It is also important to evaluate how the slide was displaced (Locat et al. 2004) and to be able to distinguish submarine landslides from sediment waves (Lee et al. 2002).

The tail of the landslide volume distribution also follows the general upper-truncated power-law form (Burroughs and Tebbens 2005), similar to the modified Gutenberg-Richter distribution for earthquake magnitudes. It is not clear whether the truncation is a result of physical limitation on maximum size of landslides or of insufficient observations of large events. However, factors (i.e., downslope length of the continental shelves) may help limit the maximum size of submarine landslides. Assumptions also need to be made while converting observed submarine landslide extent to volume.

Locat and Lee (2002) compiled a global database of known submarine landslides that included volume and run-out distance.

Slide speed and acceleration

Tsunami generation from submarine landslide is greatly affected by the time history of the slide movement: the near-field tsunami amplitude is sensitive to the initial acceleration, and the far-field tsunami is sensitive to the maximum velocity of the slide (Trifunac et al. 2002, 2003). The initial wave height may also be influenced by the depth of the center of mass of the slide (Murty 1979). There is a lack of direct observation of submarine-landslide dynamics, particularly in its tsunamigenic stages.

Landslide speed may be determined from physical modeling of landslide dynamics (Locat et al. 2004). In the absence of physical modeling, lack of direct observation limits the landslide dynamics to be modeled using a single parameter, such as its speed or duration. Likely, bathymetric slope and base friction are important parameters that affect landslide speed. Most modeling studies use speeds in the range of 66-246 ft/sec (20-75 m/sec). Maximum tsunami-wave amplitude results when the terminal speed of the slide matches the phase velocity ($C = \sqrt{gh}$) of the tsunami waves.

Cohesiveness and fluidity

The tsunami generated by a submarine landslide is also affected by the relative rigidity of the slide mass during the failure. More rigid slides that behave more like a single mass have greater efficiency at generating a tsunami than slides that quickly disintegrate into turbidity flows. Geomorphological analysis can help identify the type of initial failure, and laboratory analysis of core samples may assist in the determination of the properties of the slide material.

Investigation of tsunamigenic landslides

Investigation of tsunamigenic landslides in the area of interest may involve several steps:

1. identification of the area of interest
2. literature review of existing documentation in the area of interest
3. compilation of existing seafloor and coastal data (e.g., multibeam, seismic, cores, etc.); actual Geologic Long-Range Inclined Asdic (GLORIA, a side-scanning sonar system that produces digital images of the seafloor) images may be an excellent starting point in the U.S. Exclusive Economic Zone (EEZ) area (Schwab et al., 1993; http://coastalmap.marine.usgs.gov/aloria/). (See Section 2.4.)
4. preparation of an inventory of mass movements and corresponding slide properties (volume, shape, material properties, type of failure, etc.)
5. identification of source area
6. detailed investigation of characteristic slides, including geotechnical testing to estimate material properties
7. retrospective analysis of historical slides to estimate their tsunamigenic potential
8. detailed tsunami analysis from landslides: slope failure analysis, post-failure slide dynamics, and tsunami generation.

Subaerial landslides

The geographical areas where subaerial landslides occur are more restricted than those where submarine landslides occur. Subaerial landslides, by definition, start on land and then impact a water body. Therefore, their occurrence is generally limited to areas of steep coastal or shoreline topography. One exception to this limitation is debris flow that originates away from the shores but reaches and impacts the water body (e.g., the debris flow from 1980 collapse of the Mount St. Helens dome into Spirit Lake).

The impact velocity of subaerial landslides can be significantly greater than those for submarine landslides. However, typically, subaerial landslides displace less water than submarine slides; thus, the geographical extent of their damage is more limited.

Many of the largest subaerial landslides are triggered by earthquakes (e.g., the 1958 Lituya Bay landslide) and some by classic hillslope failure mechanisms under wet conditions (e.g., the 1963 Vaiont Reservoir landslide) (Kiersch 1964; Hendron and Patten 1985).

Tsunami generation mechanism

Subaerial landslides can be classified into two categories (Walder et al. 2003): (1) those that start near the shore and have substantial run-out under water (release-at-shore type) and (2) those that start higher up, such that a substantial portion of their run-out is over the land (initial-velocity type). The physical parameter that distinguishes these two types of subaerial landslides is the slide impact Froude number (the ratio of slide velocity at entry to the long-wave phase velocity of the tsunami waves).

The tsunami-generation mechanism for the release-on-shore subaerial landslide is very similar to that of submarine landslides. The tsunami waves are generated primarily due to the displacement of the water. On the other hand, tsunami waves from initial-velocity-type subaerial landslides are generated primarily due to the impact of the slide with the water (Heinrich 1992).

Initial wave characteristics

Due to similarities of the tsunami-generation mechanism in release-on-shore subaerial landslides and submarine landslides, these landslides can be treated similarly in terms of initial wave characteristics.

The hydrodynamics of the impact of a fast-moving mass on water is very complex. Fritz et al. (2003a, 2003b, 2004) carried out laboratory studies, and Mader and Gittings (2002) performed full, three-dimensional Navier-Stokes hydrodynamics modeling of impacts on the water body. These studies suggest that near the site of the impact, complex flow separation and crater formation occurs. Near-field wave characteristics of these strongly non-linear waves is described by Fritz et al. (2004). Lynett et al. (2006) suggested modifications to standard hydrodynamics theory for full, three-dimensional modeling of release-onshore subaerial landslide-generated waves where the slide is not completely submerged in water. Walder et al. (2003) derived scaling relationships to estimate the near-field hydrodynamic response while treating the "splash zone" as a black box because of the complexity and turbulent nature of the flow.

Source parameters

The source parameters associated with subaerial landslides are the impact Froude number and the density and the dimensions of the slide. Heinrich et al. (1998) also suggested that the frontal shape of the slide also has an effect on wave generation. It is also likely that cohesiveness of the slide material has a significant effect on wave generation.

Ice falls

An overview of the glacial processes that may result in large masses entering water is described by Richardson and Reynolds (2000). These processes include snow and ice avalanches. Ice avalanches are categorized into frontal-block failure (calving when the front of the glacier is in water), ice-slab detachment, and ice-bedrock failure (in which part of the bedrock is included in the failure).- Tinti et al. (1999) described a historical example of approximately 6 ft (2 m) waves generated in a lake by a frontal-block failure 247,000-565,000 ft^3 (7000-16,000 m^3) in volume in the western Italian Alps.

Ice avalanches moving downslope will behave similar to initial-velocity subaerial landslides, and likely primary source parameters are volume and impact Froude number. Calving will behave as a topple entering the water. Maximum volume of ice avalanches in the European Alps is given in Huggel et al. (2004).

1.3.3. Volcanoes

Tsunamis can occur due to a variety of mechanisms associated with active and Holocene-age volcanoes that are located in or near the oceans or other bodies of water. The 1883 Krakatau and mid-17th century BC Santorini tsunami events are ascribed to volcanic activity.

Source types

Beget (2000) lists the following source types associated with volcanic activity:

1. pyroclastic flow into the water
2. submarine caldera collapse
3. submarine explosion
4. debris avalanches and flank failures.

Less-common mechanisms include rapid seafloor inflation (Satake and Kanamori 1991; Kanamori et al. 1993) and coupling of the ocean with atmospheric shock waves (Beget 2000). A combination of several source types may also be a factor in the tsunami generation due to volcanic activity (e.g., 1883 Krakatau and mid-17th century BC Santorini events). No submarine volcano is continuously monitored by a volcano observatory (Begét 2000).

Pyroclastic flows

Pyroclastic flows consist of debris flows and avalanches of hot, gas-rich material that quickly flows downslope. Pyroclastic flows are produced by explosive volcanic eruptions. When the pyroclastic flow reaches the water, the hot material separates into a dust cloud of hot ash and gas and a denser material. This dense material can generate a tsunami (Legros and Druitt 2000; Watts and Waythomas 2003), and is thought to be the most efficient tsunami generator of several mechanisms investigated (Watts and Waythomas 2003).

This mechanism is similar to the subaerial landslides and avalanches. The tsunami events that have been generated by pyroclastic flow include the 1997 Montserrat (Lesser Antilles) event (Heinrich et al. 1998) and the 3500-years-before-present Aniakchak events (Waythomas and Neal 1998; Watts and Waythomas 2003). The primary source parameters, similar to the subaerial landslide source mechanism, are the impact Froude number, density, and dimensions of the tsunamigenic portion of the flow.

Submarine caldera collapse

Caldera collapses that occur under water result in a sudden depression of the water surface that hydrodynamically evolves into tsunami waves (Latter 1981; Gray and Monaghan 2003). Examples of tsunamis generated by caldera collapses are Santorini (Gray and Monaghan 2003) and the 1883 Krakatau (Nomanbhoy and Satake 1995) events, although both are thought to have generated tsunamis by a combination of volcanic mechanisms, one of which is caldera collapse. Gray and Monaghan (2003) performed laboratory experiments and numerical modeling to investigate the mechanism of wave generation and propagation caused by a caldera collapse. Their study found that the primary source parameters were caldera dimensions, fixed wall height, and the cavity depth after the collapse.

Submarine explosions

Although large submarine hydromagmatic explosions are possibly limited by large hydrostatic pressures on volcanoes (Begét 2000), regionally destructive waves caused by explosive eruptions have occurred in the past (the 1883 Krakatau event, Nomanbhoy and Satake 1995). Belousov et al. (2000) describe tsunamis generated in Karymskoye Lake in Kamchatka, Russia, by a 1996 subaquatic explosive eruption. Newhall and Self (1982) proposed the volcanic explosivity index (VEI), which is a relative measure of the size or magnitude of explosive volcanic eruptions. VEI can be linked to the potency of an explosion and, consequently, its tsunamigenic potential.

Debris avalanches andf lank failures

Volcanoes have produced some of the largest debris avalanches (Locat and Lee 2002) and, consequently, have the potential to generate severe tsunamis. Some historical examples of tsunami events generated by debris avalanches and flank failures are the 1741 Oshima, Japan (Satake and Kato 2001), the 1782 Unzen, Japan, the 1883 Augustine, the 1888 Ritter Island (Ward and Day 2003), the prehistoric Nuuanu and Wailau landslides off the Hawaiian islands (Satake et al. 2002) and off Stromboli (Tinti et al. 2000), and the 1981 Spirit Lake from Mount St. Helens dome collapse and debris avalanche (Waitt and Pierson 1994).

Avalanches can be triggered by volcanic activity or may occur during the erosional phase, but they occur most commonly from sector or flank collapses (Ui et al. 2000). Volcanic rocks commonly have greater cohesion than material that is primarily elastic, such as mud slides. This property may also increase the tsunamigenic potential of debris avalanches.

The tsunami generation from flank failures occurs through a combination of mechanisms. Volcanic spreading results in over-thrusting along a decollement on the edge of the volcanic flank (Lipman et al. 2002, 2003; Morgan et al. 2003), which results in a very shallow thrust fault similar to subduction thrusts due to earthquakes. A large earthquake on the base décollement may result in large-scale slumping on the hanging wall, such as on the south flank of Hawaii (Cannon et al. 2001). Klein et al. (2001) indicated that the volcanic flank regions of Hawaii are the most seismically active regions. Most of the far-field energy of the tsunami from the 1975 Kalapana flank earthquake was due to the earthquake; however, regionally, the concomitant landslide also added to the tsunami amplitude (Ma et al. 1999). The 1868 Kau earthquake along the southern flank of Hawaii also generated a far-field tsunami (Klein et al. 2001). Combining the tsunami displacement fields from individual mechanisms for such events may be required.

1.4. Some Historical Occurrences

This section describes some of the historical occurrences of tsunamis caused by various geophysical and seismic mechanisms. These events were selected to demonstrate several aspects of tsunami generation and analysis: (1) to illustrate the diversity of locations, including inland water bodies; (2) to illustrate the source mechanisms, including earthquakes, earthquake- induced submarine and subaerial landslides, hillslope failure, and volcanic activity; (3) to illustrate the difficulties associated with determination of the source mechanism, especially when observed data are limited; and (4) to illustrate the use of paleotsunami records in identification of the source.

1.4.1. 1958 Lituya Bay Landslide and Tsunami

Lituya Bay is located in southeastern Alaska (Miller 1960). This bay is narrow at 9 mi (14 km) long and 2 mi (3.2 km) wide at its widest point. Lituya and Crillon glaciers flow into the Gilbert and Crillon inlets, the two arms at the head of the bay, respectively. These two arms of the bay are part of a great trench along the Fairweather fault. Outer parts of the bay have gentle slopes. However, the inner parts of the bay are fiord-like with steep slopes rising 2200 ft (670 m) to 6000 ft (1828 m).

An earthquake with M_w=8.3 occurred on the Fairweather fault on July 7, 1958. The earthquake caused a large landslide on the northeast slope of the Gilbert Inlet. The landslide started at an altitude of approximately 3000 ft (914 m). The estimated volume of the landslide was more than 1 billion ft^3 (30 million m^3). The landslide caused the water in the Gilbert Inlet to surge onto the opposite slope with the runup reaching a height of 1720 ft (524 m). Miller (1960) also reported that several such large waves have occurred in Lituya Bay; the 1958 wave was the highest in terms of runup. The other four waves that occurred in 1854, 1974, 1899, and 1936 were all less than 490 ft (149 m) in terms of runup. The 1958 Lituya Bay tsunami runup is the largest documented in the NGDC tsunami runup catalog.

1.4.2. 1980 Spirit Lake Debris Flow and Tsunami

On May 18, 1980, Mount St. Helens erupted after a few-months-long activity that resulted in a bulge (or cryptodome) on the north part of the mountain. A moderate earthquake caused the entire north flank of the dome to collapse (Waitt and Pierson 1994), although some argue that the earthquake was caused by the debris avalanche (Kanamori et al. 1984). The initial landslide consisted of two separate slide-blocks that removed about 0.6 mi (1 km) of the dome. Sudden depressurization because of removal of the overburden resulted in an explosive expansion of the cryptodome, followed by a pyroclastic density current with front speed exceeding 313 mph (500 km/hr) (Waitt and Pierson 1994). Within 5 minutes of the earthquake and collapse of the flank, the density currents had spread 16 to 28 km (10 to 17.5 mi) from the source.

The toe of the pyroclastic density flow reached Spirit Lake and entered it at approximately 156 mph (250 km/hr), filling the lake with sediment and generating a tsunami by displacing the water in the lake (Waitt and Pierson 1994). The water surface of Spirit Lake was raised from 3199 ft (975 m) to 3406 ft (1038 m). The tsunami runups on the west and east arms of the lake were 820 ft (250 m) and 738 ft (225 m) above the original water surface elevation.

According to the NGDC tsunami runup catalog, the runups reported in the Spirit Lake are the second and the third largest ever observed, smaller only than the 1958 Lituya Bay event.

1.4.3. 1946 Aleutian Tsunami

On April 1, 1946, near the Aleutian trench off the Unimak Island in the eastern Aleutian Islands, an earthquake of moderate surface magnitude, M_s=7.4, generated a destructive Pacific-wide tsunami.

Researchers agree that this earthquake is one example of a tsunami earthquake, one that generates a tsunami disproportionately large compared to its surface magnitude. The tsunami generated a runup of 115 ft (35 m) on Unimak Island, Alaska, and a runup of 55 ft (16.8 m) in Hawaii. The Scotch Cap lighthouse on Unimak Island was completely destroyed and five people were killed. The tsunami resulted in 159 deaths in Hawaii, five in Alaska, two in Marquesas Islands, and one in California.

The precise tsunami-generation mechanism of this event has been a subject of much debate in the research literature. A landslide component in the generation of the tsunami was suggested by Sykes (1971) and described in detail by Kanamori (1985). The landslide component was later de-emphasized (Johnson and Satake 1997), but not conclusively ruled out. Tanioka and Seno (2001) suggested that an uplift caused by displacement of an accretionary deposit near the trench, in addition to the uplift caused by the underthrusting fault, could explain observed tsunami waveforms both in the near-field (Unimak Island) and in the far-field (Hawaii and the U.S. Pacific coastline). Fryer et al. (2004) argued that a single landslide source was responsible for this tsunami event based on the highly directive nature of the tsunami waves in the far field, rapid decay of the tsunami wave in the lateral direction, and period of the waves. Using United States Geological Survey (USGS) Geological Long Range Inclined Association imagery, they also identified a candidate landslide on the Aleutian shelf that is 15.6 mi (25 km) wide, 40.6 mi (65 km) long, and has a volume of 48.8-73.2 mi^3 (200-300 km^3). López and Okal (2006) used more recently available seismic data and concluded that a very slow rupture accounts for the large far-field tsunami amplitudes, and a concomitant landslide is necessary to explain the large near-field runup at Scotch Cap.

1.4.4. 1929 Grand Banks Landslide and Tsunami

On November 18, 1929, an earthquake of $M_w = 7.4$ occurred on the southern edge of the Grand Banks, about 175 mi (280 km) south of Newfoundland. The Grand Banks are a group of underwater plateaus off the coast of Newfoundland on the North American continental shelf. The earthquake caused a large landslide that turned into a turbidity current, flowing more than 625 mi (1000 km) to the east (Fine et al. 2005). The turbidity current severely damaged trans-Atlantic telegraph cables. Although all of the cables along the continental slopes were broken, none on the continental shelf was damaged.

The area of the submarine landslide coincided with the epicenter of the earthquake. Later, ocean-floor mappings and timings of the cable breaks were used to estimate the area of the landslide at approximately 7813 mi^2 (20,000 km^2). The estimated volume of this landslide was 48.8 mi^3 (200 km^3). The landslide generated a trans-Atlantic tsunami that was observed as far as the Azores Islands and Portugal. The tsunami was also recorded on tide gauges at Atlantic City, New Jersey, and Charleston, South Carolina. The tsunami resulted in 27 deaths in Newfoundland and one in Nova Scotia.

The runups reported in the NGDC tsunami runup database for the 1929 Grand Banks tsunami were 23 ft (7 m) in Taylor's Bay, 15.4 ft (4.7 m) in Placentia Bay, and 15 ft (4.57 m) at Burin and Port Au Bras, all in Newfoundland. On the U.S. Atlantic coast, recorded runups were a foot (0.3 m) at Ocean City, Maryland, and 2.2 ft (0.68 m) at Atlantic City, New Jersey. The 1929 Grand Banks tsunami is notable for a couple of reasons: it was one of the very few catastrophic tsunamis to occur in the Atlantic, and it was one of the very few transoceanic tsunamis generated by a landslide.

1.4.5. 1964 Valdez Arm Landslide and Tsunami

On March 27, 1964, the second-largest earthquake on record, with M_w = 9.2, occurred in Prince William Sound in Alaska at the boundary of the Pacific plate, subducting beneath the North American plate. The epicenter was approximately 6.25 mi (10 km) east of the mouth of College Fjord, and approximately 56 mi (90 km) west of the town of Valdez (old Valdez). The depth of the earthquake was about 15.6 mi (25 km). The area affected by the vertical deformation due to the earthquake was estimated to be approximately 100,000 mi^2 (250,000 km^2). The average slip was estimated to be 14.4 ft (9 m).

The great earthquake resulted in several subaerial and submarine landslides on and near the Alaskan coast. The town of Valdez was built on unconsolidated deltaic deposits. During the earthquake, the sediments under the waterfront area near the town liquefied under seismic vibrations, and a large section of the delta, approximately 4000 ft (1219 m) long by 600 ft (183 m) wide, collapsed into the bay and generated a local tsunami. The combined effects of the earthquake and the local tsunami destroyed the waterfront, including the facilities of the Valdez port. The earthquake also generated a tsunami that reached Valdez several hours later. The runups from the local tsunami were 170 ft (51.8 m) at the Valdez Inlet, 113 ft (34.4 m) at Kings Bay, 100 ft (30.5 m) at Aialik, and 79 ft (24.2 m) at Blackstone Bay, all in Alaska. The maximum runups due to the trans- Pacific tsunami generated by the earthquake in Washington State was 15 ft (4.5 m) at Wreck Creek; that in Oregon was 11.5 ft (3.5 m) at Bandon; at Depoe Bay, Nehalem River, and at Yaquina Bay, that in California was 28.9 ft (8.8 m) at Van Damme State Park, and that in Hawaii was 16 ft (4.9 m) in Waimea Bay, Oahu, Hawaii. The damage at the old Valdez town was so great that the town was moved from its old location to a new area. The new town is located at a higher elevation and on more-stable ground.

The earthquake also caused major damage in Seward, Alaska, at the head of Resurrection Bay, along the east side of the Kenai Peninsula. An approximately 0.6-mi- (1-km-) long section of the waterfront began sliding toward the sea, caused by large-scale offshore landslides, within one minute of the strong ground shaking. This area contained three large docks, many oil tanks, and a railroad yard. Pipes from the oil tanks ruptured and the oil tanks overturned causing a fire. The offshore landslides generated a tsunami 32.8 ft (10 m) in height that struck south of Seward, causing extensive damage. Seismically generated tsunami waves, also 32.8 ft (10 m) high arrived approximately 30 minutes later, and caused additional damage. In all, 13 deaths in were reported in Seward. The NGDC tsunami runup database lists a runup of 27.2 ft (8.3 m) for Seward from the earthquake and landslide-generated tsunami.

This event is one example in which the earthquake ground motion occurred coincident with the local tsunami generated from the subaerial landslide into the bay. The damage, therefore, was a combined effect of both earthquake ground motion and the flood wave.

1.4.6. 1960 Chile Earthquake and Tsunami

On May 22, 1960, the largest recorded earthquake, of M_w=9.5, occurred off the coast of south-central Chile, in the Nazca subduction zone, where the Nazca plate is subsiding below the South Americal plate. The total rupture length, over a period of days, was approximately 625 mi (1000 km), one of the longest ever reported. The earthquake gencrated a trans-Pacific tsunami. The tsunami caused 61 deaths and $75 million in damages in Hawaii. In Japan, the tsunami resulted in 138 deaths and $50 million in damages. Thirty-two deaths were reported in the Philippines from the tsunami. The U.S. Pacific Coast sustained $500 million damage. The near-field tsunami devastated the Chilean coast. More than 2000 deaths, 3000 injuries, 2 million homeless, and $550 million in damages were reported in southern Chile.

The earthquake caused widespread land deformation along much of the fault. A subsidence of approximately 6.6 ft (2 m) occurred in the coastal mountains; uplifts of 1.6 ft (0.5 m) were observed along the foothills of the Andes, and offshore islands were raised up to 19.7 ft (6 m). The Puyehue volcano erupted two days after the earthquake. Several landslides near Tralcan mountains blocked San Pedro River that drains Rifiihue Lake, the lowest of the seven lakes that receive inflow from Enco River. Due to the blockage, water in Rifiihue Lake started to rise rapidly and would have resulted in a catastrophic flood downstream if it had overtopped the 78.7-ft-(24-m-) high dam, affecting several towns and the city of Valdivia. An effort was undertaken by the military, an electric utility company, and an economic-development organization of the government to control the waters in Rinihue Lake. Several drainages to the other lakes were dammed to reduce inflow into the lake chain. The main dam was lowered from 78.7 ft (24 m) to 49.2 **ft** (15 m) to allow approximately 106 billion ft^3 (3 billion m^3) of water to drain from the lake gradually. The work took two months to complete. For this event, the NGDC database reported the following tsunami runups shown in Table 1-1.

Table 1-1. Runups Recorded at Various Locations for the 1960 Chile Tsunami

Location	Maximum Runup
Australia	5.6 ft (1.7 m)
Chile	82 ft (25 m)
Japan	21 ft (6.4 m)
Mexico	8.2 ft (2.5 m)
New Zealand	13.1 ft (4 m)
Papua New Guinea	5.9 ft(1.8 m)
Peru	5.6 ft (1.7 m)
Pitcairn Islands	40 ft (12.2 m)
Russia	15.4 ft (4.7 m)
Samoa	16 ft (4.9 m)
United States	Up to 35 ft (10.7 m)
Montague Island, Alaska	7.5 ft (2.29 m)
Cresent City, California	5.5 ft (1.68 in)
Princeton, California	7.25 ft (2.21 m)
Stinson Beach, California	5 ft (1.52 tn)
Ahihi Bay, Maui, Hawaii	8.9 ft (2.7 na)
Anakua Point, Kauai, Hawaii	9.8 ft (3 in)
Aweoweonui, Hawaii	10.8 ft (3.3 in)
Coconut Island, Hilo, Hawaii	15 ft (4.6 in)
Haena, Kauai, Hawaii	13.5 ft (4.1 m)
Hana Bay, Maui, Hawaii	15 ft (4.6 m)
Hilo, Hawaii	35 ft (10.7m)
Honolulu Landing, Oahu, Hawaii	14.1 ft (4.3 m)
Honuapo, Hawaii	17 ft (5.2 m)
Kaalualu Bay, Hawaii	17 ft (5.2 m)
Southeastern coast of Hawaii	17 ft (5.2 in)
Depoe Bay, Oregon	5.9 ft (1.8 m)
Seaside, Oregon	4.9 ft (1.5 m)
Willapa Bay, Washington	2 ft (0.61 in)

1.4.7. The "Orphan" Tsunami of 1700

Accounts of flooding from a tsunami in January, 1700, exist in Japanese documents written at that time (Atwater et al. 2005). The described effects of the tsunami include villagers fleeing to higher ground, damage to salt kilns and fishing shacks, drowned crops, water flooding a castle moat and entering a government storehouse, houses and buildings that were washed away, and a fire. There was no advance warning of an earthquake preceding the tsunami. In fact, no account mentioned associated shaking, and two accounts noted the lack of an earthquake warning. By the 1990s, this tsunami had become Japan's best- documented tsunami of unknown origin. The runups from the 1700 tsunami in Japan were estimated to range from about 6.6 ft (2 m) to more than 16.4 ft (5 m) (Satake et al. 2003; Tsuji et al. 1998) on the eastern Japanese coast.

Because there was no mention of a nearby earthquake in the Japanese documents for the 1700 tsunami, researchers in the 1990s attempted to find its source far away from Japan. Potential sources around the Pacific Rim, other than Cascadia, conflicted either with the year of occurrence of the tsunami or with the wave heights and runups estimated for the tsunami. South American catalogs were used to identify sources of tsunamis observed in Japan in 1687 (Callao earthquake), 1730 (Valparaiso earthquake), and 1751 (Concepción earthquake). The 1952 Kamchatka earthquake, which was the third largest of the 20th century, produced wave heights comparable to the 1700 tsunami only in the north of Japan. The second largest earthquake of the 20th century, the 1964 Alaska earthquake that produced thc devastating tsunami in Valdez Arm of Prince William Sound, had a strong directivity away from Japan and produced wave heights of less than 3 ft (1 m) in Japan. The Cascadia source for the 1700 tsunami was proposed by Satake et al. (1996), who not only were able to find evidence to support the date of occurrence of this great source earthquake, but also, remarkably, the hour. Satake et al. (1996) also estimated that the moment magnitude of this Cascadia earthquake was most likely approximately 9. They presented arguments in favor of this opinion based on the estimated wave heights on the coast of Japan for the 1700 tsunami, and based on numerical simulations of a trans- Pacific tsunami with the source in Cascadia. Also, an earthquake of M_w=8.2 that occurred on October 4, 1994, off the Kuril Islands near Japan produced tsunami heights less than 1 ft (0.3 m) on the Pacific Coast of the United States. Satake et al. (1996) suggested that, reciprocally, a source in Cascadia that produced wave heights greater than 6.6 ft (2 m) on the east coast of Japan would have to be significantly larger than M_w=8.2.

As the study of plate tectonics and giant earthquakes (those of $M_w>9$) advanced during the 1960s through the 1980s, much debate took place regarding the nature of seismicity in the Cascadia. Efforts were undertaken in the late 1980s to identify geological evidence of past giant earthquakes in the area (Atwater 1987).

Subduction zones are characterized by down- warping of the subducting place and bulging of the overriding plate as stress builds up at the interplate boundary. During earthquakes, the leading edge of the overriding plate breaks free, releasing the stress; the seaward portion of the overriding plate springs up and the continental side subsides. The effect of the seaward, upward springing creates the leading tsunami wave by upward displacement of the water column. The leading positive wave travels away from the continental side of the subduction zone. On the continental side, subsidence results in an initial depression of the water column that is frequently observed as the initial withdrawal of the sea before the arrival of the first (positive) tsunami wave. The subsidence on the coast results in previously dry coastal land becoming submerged by sea water in a very short period of time. The effects of a tsunami that follow the earthquake may also lead to deposition of sand or mud layers. In the late 1980s, geologists investigated bays and river mouths along the Pacific Coast. At nearly every site, they found evidence that land had dropped. The evidence included groves of dead tree trunks of western red cedar that stand in tidal marshes, termed ghost forests. Also, thousands of tree stumps buried in the marshes were seen exposed by flowing water in the banks of tidal streams. The investigation also found buried remains of tidal marshes. In sediment cores, investigators found muddy tidal deposits on peaty marsh soils, which supports evidence of past earthquakes in Cascadia.

Geologists in the 1980s and 1990s also found evidence of tsunamis that consisted of sand sheets near bays and mouths of rivers. The sand sheets taper inland and contain microscopic siliceous shells of marine diatoms. In alternating layers of sand and mud, the researchers found that, at most sites, sand arrived just before tidal mud began covering freshly subsided soil. This evidence led them to conclude that the sand was brought in by a tsunami of nearby origin. The sequence of events put forth was that an earthquake on the interplate boundary abruptly generated a local tsunami and, at the same time, lowered the adjoining coast; the subsequent near-field tsunami created sand deposits. Using radiocarbon dating and tree-ring interpretation, researchers estimated that nearly all sites from southern British Columbia to northern California along the Pacific Coast had most recently subsided within the past

400 or 500 years, and that the southern Washington coast had subsided after 1680 (Atwater et al. 2005).

Researchers also attempted to detect differences in timing of the earthquake-induced subsidence among the sites. Differences in timing would indicate different source earthquakes and would also limit the size of the ruptures. Radiocarbon dating of dead plants and tree-ring pattern matching did not narrow the timing to more than a few decades. The evidence did not conclusively prove either hypothesis of Cascadia rupture; a single giant rupture and earthquake that would require all dates to coincide or a series of smaller earthquakes during a few decades. However, the time window of either of these two Cascadia events was indeed reduced to the period of 1695-1720. The research of Satake et al. (1996) fmally brought the evidence from Cascadia together with the written records from Japan to conclude that the parent earthquake of the 1700 orphan tsunami was a giant earthquake in Cascadia.

The evolution of our understanding with respect to the 1700 tsunami is an excellent example of paleotsunami research complemented by old records and anecdotal evidence.

1.4.8. 1963 Vaiont Reservoir Landslide and Tsunami

On October 9, 1963, a massive landslide occurred on the southern hillslope of the reservoir behind Vaiont Dam, located about 100 mi from Venice, Italy (Kiersch 1964). The landslide was caused by shear failure of the overlying limestone along thin clay layers under increased pore pressure. The landslide area was 1.25 mi (2 km) long and 1 mi (1.6 km) wide. The total estimated volume of the slide was 8.4 billion ft^3 (238 million m^3).

The landslide filled the reservoir 1.25 mi (2 km) upstream from the dam completely to a height of 574 ft (175 m) above the original water level. The landslide speed was reported as reaching 50-100 ft/sec (15-30 m/s). The tsunami wave generated by the slide ran up more than 850 ft (260 m) above the original reservoir water level. The tsunami wave also overtopped the dam by an estimated 328 ft (100 m) and flowed downstream. The height of the flood wave approximately 1 mi (1.6 km) downstream from the dam was estimated to be 230 ft (70 m).

1.4.9. 2004 Sumatra Earthquake and Tsunami

On December 26, 2004, the fourth largest recorded earthquake of M_w=9.1 occurred off the western coast of northern Sumatra in the Indian Ocean. Stein and Okal (2007) presented an analysis that may revise the moment magnitude of this earthquake to M_w=9.3 , making it the second largest on record, after the great 1960 earthquake in Chile. The earthquake occurred due to the subduction of the Indian plate below the Burma microplate, which is a sliver plate between the Indian and the Sunda plates. This earthquake was the first giant (M_w>9) earthquake since the 1964 Alaska earthquake.

The USGS reported the epicenter at 3.3° N, approximately 155 mi (255 km) south-southeast of Band Aceh, Sumatra, Indonesia. The rupture progressed north to about 7° N, with a total rupture length of approximately 750 mi (1200 km), and the rupture zone was reported to be approximately 125 mi (200 km) wide with an average slip of 36 ft (11 m) (Stein and Okal 2007).

This earthquake generated a devastating transoceanic tsunami that traveled across the globe. The runup measured in Crescent City, California, was 2 ft (0.61 m); that at Atlantic City, New Jersey, was 0.75 ft (0.23 m); and that at Trident Pier, Florida, was 1.1 ft (0.34 m). The near-field tsunami in Sumatra devastated the coastline, with a maximum runup of 167 ft (50.9 m) at Labuhan on the northwest coast of Sumatra. The tsunami traveled across the Indian Ocean and affected several countries, including India [maximum runup of 31.4 ft (9.56 m)], Malaysia [maximum runup of 13.1 ft (4 m)], the Maldives [maximum runup of 14.5 ft (4.43 m)], Myanmar [maximum runup of 9.5 ft (2.9 m)], Somalia [maximum runup of 31.2 ft(9.5 m)], Sri Lanka [maximum runup of 37 ft (11.3 m)], and Thailand [maximum runup of 64.2 ft (19.57 m)]. The tsunami caused approximately 300,000 deaths and unprecedented damage in countries around the Indian Ocean, with Indonesia accounting for more than 228,000 fatalities.

The Madras Atomic Power Station, a commercial nuclear power plant owned and operated by the Nuclear Power Corporation of India Limited (NPCIL), and located near Chennai, India, was affected by the tsunami generated by the 2004 Sumatra earthquake (NPCIL 2007). At approximately 9:15 pm local time on December 26, 2004, the condenser cooling pumps of Unit 2 of the installation were affected due to flooding of the pump house and subsequent submergence of the seawater pumps by tsunami waves. The turbine was tripped and the reactor shut down. The unit was brought to a cold-shutdown state, and the shutdown-cooling systems were reported as operating safely. Unit 1 of the installation was shut down for refurbishment at the time of this event. After detailed inspection of the plant by the NPCIL and the Atomic Energy Regulatory Board of India, the plant resumed normal operations on January 2, 2005. Adjacent to the Madras Atomic Power Station, a prototype fast-breeder reactor was under construction at the time of the 2004 tsunami. Work on the concrete raft at the bottom of the excavated pit was in progress when the tsunami waves flooded the pit. Most of the workers were safely evacuated due to the alertness of supervisors; however, there was one fatality. The tsunami waves deposited debris and sediment in the pit over the partially poured concrete. The pit was subsequently dewatered, and the debris was removed before construction resumed.

1.5. Landslides in Earth's Oceans

The understanding of submarine landslides is a recent development spurred by availability of high-quality seafloor mapping data [e.g., the USGS GLORIA mapping effort of the USGS started in 1984]. GLORIA is a side-scanning sonar system that produces digital images of the seafloor. The intensity of backscattered sound is a function of the gradient, the surface roughness, and the texture of the seafloor. Raw GLORIA data were processed to correct for geometric and radiometric distortions and mosiacked to produce 2° by 2° images of the survey area. Sixteen images were produced for the Gulf of Mexico, 21 for the Atlantic coast, and 36 for the Pacific Coast.

Another tool for mapping the bathymetry of the seafloor is the use of multibeam sonar system (Hughes-Clarke et al. 1996). Multibeam sonars first became available in 1971, but hardware to, deal with the large volumes of data produced by these systems and the limitations of positioning systems available at the time precluded their general use. Since then, many such systems have become available and rapid processing of the data advanced. Initially, these systems were used for industrial applications, such as surveys of offshore platform sites and pipeline route corridors. The values of this tool for bathymetric surveys was recognized with time.

GLORIA imagery has been used to identify submarine landslides near the Hawaiian islands (Lipman et al. 1988; Moore et al. 1989, 1994a, b), mapping of features of the continental slope in the Gulf of Mexico (Rothwell et al. 1991) and landslides (Twichell et al. 1993), and investigation of landslide zones in the Atlantic (Booth et al. 1993; McGregor et al. 1993; O'Leary et al. 1987; O'Leary 1993; Popenoe et al. 1993; Schwab et al. 1991, 1993).

1.5.1. The Pacific Ocean

McAdoo et al. (2000) present results of a morphometric analysis of submarine landslides off the coasts of Oregon, California, Texas, and New Jersey. They used multibeam bathymetry data, in addition to GLORIA images, to identify scars left by submarine landslides. Widespread evidence of slope failure and submarine landslides was found in each of the four margins investigated. Properties of these landslides were measured or estimated using multibeam bathymetry and GLORIA images. The parameters of the landslides measured by the authors included latitude, longitude, area, depth of the headscarp of the landslide, and runout length of the landslide. The volumes of the landslides were estimated using the thickness of the slides and assuming a wedge geometry for the slides. The authors presented a table of parameters for the investigated submarine landslides (see Table 1 of McAdoo et al. 2000).

This table included 20 submarine landslides in the Oregon margin with slide areas ranging from 0.4 mi^2 (1 km^2) to 93.4 mi^2 (239 km^2). Depth to the headscarp for these slides ranged from 1800 ft (549 m) to nearly 9600 ft (2924 m). The estimated volumes for the Oregon margin landslides ranged from less than 0.02 mi^3, or 3.6 billion ft^3 (0.1 km^3), to 10.4 mi^3, or 1527.3 billion ft^3 (42.5 km^3). The runout lengths varied from less than 0.6 mi (1 km) to 13.8 mi (22 km).

In the California margin, 25 submarine landslides were identified by McAdoo et al. (2000) that ranged in area from 2.7 mi^2 (6.8 km^2) to 205 mi^2 (525 km^2). Depth to the headscarp of these slides ranged from 2707 ft (825 m) to 8914 ft (2717 m). The estimated volumes of the landslides ranged from 0.05 mi3, or 7.2 billion ft(0.2 km^3), to 6.7 mi3, or 991.9 billion ft^3 (27.6 km^3). The runout lengths varied from 1.7 mi (2.7 km) to 31.9 mi (51 km).

Mass movement features in the Monterey Bay region of the central California coast are described by Greene and Ward (2003). Multibeam bathymetric data collected by the Monterey Bay Aquarium Research Institute show mass wasting occurring along the northern slope of the Santa Barbara basin. Similar data collected by the USGS offshore from Long Beach, California, show a large debris avalanche. A small landslide was reported that occurred at the head of the Monterey Canyon during the 1989 Loma Prieta earthquake. A small tsunami, approximately 1.6 ft (0.5 m) high, was generated from this event. Extensive mass wasting, including larger landslides, covering an area of 184 mi^2 (470 km^2), are identified on the continental slope of the Monterey Bay region. Individual landslides range in size from small 0.4 mi^2 (1 km^2) to very large—comparable to the size of the Monterey Bay—and exceed 8.5 mi^3 or 1258 billion ft^3 (35 km^3) in volume.

Greene and Ward (2003) reported simulated tsunamis from three identified landslides. The Ascension Canyon landslide is located at a depth of approximately 1969 ft (600 m), is 0.94 mi (1.5 km) long, 1214 ft (370 m) wide, and about 164 ft (50 m) thick. According to Greene and Ward (2003) modeling, the initially generated tsunami wave was nearly 6.6 ft (2 m) high.

The runup was calculated to be 9.8 ft (2.98 m) on the coast near Davenport, well below the 32.8 ft (10 m) high coastal cliffs. The runup on the Monterey Peninsula was estimated to be 1.7 ft (0.52 m), and those at Santa Cruz and Moss Landing were estimated to be 4.4 ft (1.34 m) and 1.2 ft (0.36 m).

The second landslide used by Green and Ward (2003) to simulate a tsunami was the Tubeworm Slump in the Monterey Canyon. The slide area was 6.3 mi^2 (16 km^2) with a runout of 984 ft (300 m) and a volume of 0.6 mi^3 or 86 billion ft^3 (2.4 km^3). The first arrival of the tsunami waves at the Monterey Peninsula was a 3.3-ft (1-m) high positive wave, and that at Santa Cruz was a 6.6-ft (2-m) deep trough.

The third tsunamigenic landslide used by Greene and Ward (2003) was a large mass-movement field exceeding 35 mi^2 (90 km^2) in area in the Monterey Meander. The mass movement is composed of several retrogressive slumps and debris flows. In the event of a hypothetical large failure in this area, the authors estimated that tsunami waves could reach heights of 36.1 ft (11 m) in the Monterey area, 26.2 ft (8 m) at Moss Landing, and 6.6 ft (2 m) at Santa Cruz.

Lee et al. (2003) describe case studies of three submarine landslide environments: Resurrection Bay, Alaska; Commencement Bay, Washington; and the Los Angeles margin near Palos Verdes, California. These submarine landslides and mass- movement sites were identified on recently available multibeam-sonar data.

The Resurrection Bay landslide, which occurred during the great 1964 Alaska earthquake, generated local tsunamis that devastated the waterfront at Seward, Alaska (see Section 1.4.5).

Commencement Bay is located near Tacoma, Washington. The Puyallup River flows into the bay, carrying meltwater from the glaciers of Mount Rainier, including large quantities of glacial silt leading to the growth of a delta at an average rate of 1.6 mi (2.5 km) per 1000 years (Dragovich et al. 1994). The high sediment- accumulation rate leads to high excess pore-water pressure, creating favorable conditions for slope failure. Two failures have been recorded in the past (Gardner et al. 2001), one in 1894, without a precursor earthquake, and another in 1992 during the Nisqually earthquake. The 1894 failure resulted in a tsunami reported to be 10 to 20 ft (3 to 6 m) high. The 1992 event did not cause any tsunami. Lee et al. (2003) identified a number of features on the multibeam bathymetry map of Commencement Bay, including large sediment lobes off the mouth of the Puyallup River. They also mapped a 1.25-mi (2-km) long submarine channel and identified it as the 1894 landslide. The channel is 1476 ft (450 m) wide at its head.

The submarine-landslide field located to the south of the Palos Verdes Peninsula has been known for a long time, but the details of the morphology became available recently using multibeam bathymetric surveys (Gardner and Mayer 1998; Gardner et al. 1999). Two landslides were identified on the multibeam imagery. One of the landslides was dated using radiocarbon techniques and found to be approximately 7500 years old. The age of the other landslide is unknown. Landslide debris was identified as far as 6.3 mi (10 km) from the base of the slope, indicating the landslides had considerable momentum. Lee et al. (2003) concluded that both landslides were probably tsunamigenic. Locat et al. (2004) developed a simplified model of the landslide and simulated a tsunami based on this event. The amplitude of the tsunami was not estimated precisely because of uncertainty in landslide source parameters.

1.5.2. The Gulf of Mexico

In the Gulf of Mexico, 25 submarine landslides were identified by McAdoo et al. (2000) with areas ranging from 3.8 mi^2 (9.6 km^2) to 2152 mi^2 (5509 km^2). Depth to the headscarp of these landslides varied from 397 ft (121 m) to 7871 ft (2399 m). The estimated volume of these landslides ranged from 0.2 mi^3, or 28.7 billion ft^3 (0.8 km^3), to 37.2 mi^3, or 5470 billion ft^3 (152.2 km^3). The runout lengths varied from 1.9 mi (3 km) to 104 mi (167 km). Tripsanas el al. (2003) described the slope instabilities in the Bryant Canyon area in the northwest Gulf of Mexico. Bryant Canyon acts as a conduit for the transport of sediment from the Mississippi River to the abyssal plain of the Gulf of Mexico.

Ten Brink et al. (2008) conducted a recent literature survey to identify landslide sources in the Gulf of Mexico. They described three distinct geologic provinces in the Gulf of Mexico Basin: a carbonate province, a salt province, and a canyon-to-deepsea-fan province. Salt that originally underlay Louisiana, southern Texas, and offshore of the Bay of Campeche in Mexico was eroded and subsequently deposited during the early development of the basin. Subsequently, sediment eroded off the North American continent was deposited over the salt and the overburden caused the salt to migrate seaward. At present, the salt underlies large parts of the northern Gulf of Mexico and the southwest corner of the Gulf of Mexico in the Bay of Campeche. The Sigsbee Escarpment is a cliff located south of Louisiana and Texas that marks the seaward limit of the salt. The deposition of the salt was followed by formation of carbonate reefs along much of the margin of the Gulf of Mexico Basin. At present, the reef system is exposed along the Florida Escarpment and the Campeche Escarpment. The escarpments stand approximately 4900 ft (1500 m) above the abyssal plain with gradients locally as steep as 20°. Sediment from erosion by North American rivers delivered sediment into the basin that resulted in the creation of a series of deep-sea fans Three of these fans are the Bryant Fan, Mississippi Fan, and Eastern Mississippi Fan. The Mississippi Fan is the largest and covers most of the eastern part of the gulf.

Landslides have occurred in each of the three provinces of the Gulf of Mexico Basin (ten Brink et al. 2008). Landslides in the carbonate province have occurred both on the steep Florida and Campeche Escarpments as well as on gentler slopes above the escarpments. A large talus deposit[(a)] was identified at the base of the Campeche Escarpment, but its full extent and the amount of material of an individual failure is unknown. Talus blocks are also observed along the base of the southern Florida Escarpment. Talus deposits at heads of some box canyons are less than 6 mi^2 (15 km^2) in area. Large collapse scars along the central part of the West Florida Slope are present. The entire slide scar is approximately 75 mi (120 km) long and 19 mi (30 km) wide and may have a total volume of 244 mi^3 (1000 km^3). The slide, located at approximately the same latitude as Tampa, Florida, may have had at least three generations of failure. Another extensive area of collapse was mapped in the southern portion of the West Florida slope that may be 6-11 mi (10-17 km) long.

Ten Brink et al. (2008) did not find any published reference to landslides in the salt province in the Bay of Campeche. Detailed bathymetric mapping of the northern Gulf of Mexico revealed a unique morphology consisting of relatively small circular basin features 2.5-20 mi (4-33 km) wide with areas of 2-122 mi^2 (5-312 km^2). The GLORIA imagery has identified 37 landslides in the salt province and along the base of the Sigsbee Escarpment. The largest of these landslides is located in the northwest part of the Gulf of Mexico and is approximately 71 mi (114 km) long, 33 mi (53 km) wide, and 879 mi^2 (2250 km^2) in area.

A large landslide complex, approximately 9000 mi^2 (23,000 km^2) in area, was identified in the upper portions of the Mississippi Fan (ten Brink et al. 2008). The total volume of the deposits have not been estimated accurately. The Eastern Mississippi Fan also has a relatively large landslide that partially buries the canyon at the head of the fan and is approximately 96 mi (154 km) long, 14 mi (22 km) wide, and 941 mi^2 (2410 km^2) in area.

[(a)] Sloping mass of rock debris deposited at the base of a cliff.

The characteristics of submarine landslides in the Gulf of Mexico are not well understood due to a lack of data in certain regions. The age, style, and distribution of landslides is still incomplete. Ten Brink et al. (2008) recommended that age dating, compilation of multibeam bathymetry data, and mapping of the Campeche Bay area are required to refine understanding of submarine landslides in the Gulf of Mexico. Until further research is undertaken and completed, tsunami-hazard assessment for landslide sources in the Gulf of Mexico should be carried out on a case-by-case basis.

1.5.3. The Atlantic Ocean

In the Atlantic Ocean, a giant submarine landslide complex known as the Storegga landslide complex, (Bryn et al. 2003), is located off the coast of Norway. The slide is gigantic, 854 mi^3 or 126 trillion ft^3 (3500 km^3) in volume, 35,156 mi^2 (90,000 km^2) in area, 10,547 mi^2 (27,000 km^2) in slide-scar area, with 281 mi (450 km) long runout. The slide is characterized by several strata of buried mass movements that occur on parallel slip surfaces in marine clay layers (Bryn et al. 2003). These slides have occurred on a semi-regular basis during the past 500,000 years, in good agreement with the cycles of main continental-shelf glaciation. The trigger mechanism is believed to be seismic activity associated with glacioisostatic rebound.[(a)] The most recent of the Storegga slides, believed to have occurred around 8200 years ago, also generated a tsunami that reached surrounding coasts (Bondevik et al. 1997).

In the Faroe Shetland Channel, located between the Faroe Islands and Shetland Islands in the Atlantic, a slide known as the Afen slide was first recognized on Towed Ocean Bottom Instrument (TOBI) side-scanning sonar data. The data was acquired for the Atlantic Frontiers Environmental Network (AFEN). The maximum width of this slide is 1.9 mi (3 km), and the length of the scour and debris lobe is 7.5 mi (12 km) (Bulat 2003). Wilson et al. (2003) used the TOBI image compiled by Bulat (2003), the British Geological Survey regional seismic survey data, and geotechnical data obtained from sediment cores within the slide area to assess phases, modes, and characteristics of the Afen slide.

In the continental margin off the coast of New Jersey, McAdoo et al. (2000) identified 13 submarine landslides that ranged in area from 2 mi^2 (5 km^2) to 24.2 mi^2 (62 km^2). Depth to the headscarp of these landslides varied from 4675 ft (1425 m) to 7051 ft (2149 m). The estimated volume of these landslides ranged from 0.07 mi^3, or 10.8 billion ft^3 (0.3 km^3), to 1.3 mi^3, or

[(a)] The uplifting of land once depressed under the weight of glacial ice (Ranalli, 2001).

187 billion ft^3 (5.2 km^3). The runout lengths varied from 1.6 mi (2.5 km) to 13.1 mi (21 km).

Ward and Day (2001) described the potential collapse of the Cumbre Vieja volcano on the island of La Palma which is in the Canary Islands, off the west coast of Morocco. The Cumbre Vieja volcano has been the most active volcano in the Canary Islands during the last 125,000 years. It occupies the southern one-third of the La Palma Island, with slopes of 15 to 20 degrees. The flank- failure scenario described by Ward and Day (2001) involves a single N-S rift under the west flank of the volcano. The detachment fault[(a)] surfaced as a west-dipping normal fault along the crest of the volcano during its 1949 eruption. The scrap extends 2.5 mi (4 km) and has a maximum offset of 13.1 ft (4 m). The fault has been inactive since 1949. Ward and Day (2001) postulated that a future eruption near the top of the Cumbre Vieja will trigger a flank failure. Based on geological evidence from existing lateral-collapse scars, bathymetric and imaging sonar surveys of La Palma and other locations, and comparisons with other flank failures of volcanoes (Mount St. Helens), the authors estimated the future dimensions of the slide block could be 9.4-12.5 mi (15-20 km) in width and 9.4-15.6 mi (15-25 km) in length. The authors suggested that the mean thickness of the slide block could range from 0.6 mi (1 km) to 1.2 mi (2 km). Based on these dimensions, Ward and Day (2001) estimated that the volume of a future flank failure could range from 36.6 mi^3, or 5391 billion ft^3 (150 km^3), to 122.1 mi^3, or 17,969 billion ft^3 (500 km^3).

Ward and Day (2001) considered the worst-case slide block of 122.1 mi^3 or 17,969 billion ft^3 (500 km^3) in volume that break away and falls into the ocean and slide for 37.5 mi (60 km) until it reaches flat ocean bottom at a depth of 13,123 ft (4000 m). The authors proposed a maximum runout velocity of 328 ftls (100 m/s). The authors noted that the chosen runout velocity was substantially less than the tsunami celerity, a minimum of approximately 459 ft/s (140 m/s), corresponding to the ocean depth near the La Palma slide in excess of 6562 ft (2000 m). The tsunami simulations carried out by Ward and Day (2001) resulted in a 2953-ft (900-m) high dome of water representing the initial waveform, slightly less than the mean thickness of the sliding block. The simulation resulted in 32.8-ft (10-m) high waves in Newfoundland and those of 65.6-82 ft (20-25 m) height on the Florida coast.

Mader (2001) described another attempt to model the tsunami generated from a worst-case flank failure of the Cumbre Vieja volcano. Physical modeling of the slide moving as a single block by Herman Fritz, using the

---

[(a)] Parallel zones of weakness inside a volcano along which subsidence often occurs.

experimental setup described by Fritz et al. (2001), indicated an initial waveform with a 2133 ft (650 m) waveheight, a 18.8-25 mi (30-40 km) wavelength, and a 3-4 minute period. The wave was described as an intermediate wave and not a shallow-water tsunami wave. The wave would disperse significantly as it propagated away from the source in the Atlantic. Mader (2001) simulated the tsunami across the Atlantic using the nonlinear shallow- water model SWAN (Mader 2004) that includes Coriolis and frictional effects. The author noted that using a shallow-water wave model to describe the propagation of intermediate waves would only provide an estimate of the upper limit of the amplitude and the period of the waves. A 10-minute Atlantic bathymetry generated from the 2-minute Mercator Global Marine Gravity topography of the earth was used in the tsunami simulation. The initial waveform was a 2133 ft (650 m) high, with a 12.5 mi (20 km) radius. The maximum simulated tsunami wave heights at deep-water locations off the eastern coast of the United States was 9.8 ft (3.0 m) at a water depth of 9587 ft (2922 m) east of Washington, D.C.

Mader (2001) also estimated the effect of dispersion on the propagation of the tsunami caused by the Cumbre Vieja flank failure using the ZUNI model (Mader 2004) that solves two- dimensional time-dependent Navier-Stokes equations for incompressible flow. The author concluded, based on the results of the ZUNI simulations, that the tsunami wave heights off the coast of the United States would be less than 1 m and, even after amplification of the tsunami waves due to shoaling, the waves would not present a major hazard.

Pararas-Carayannis (2002) critically evaluated the Cumbre Vieja flank-failure tsunami-simulation studies regarding three aspects: (1) if the massive volcanic collapses could occur as postulated, (2) if the source dimensions and initial modeling parameters were correctly evaluated from proper coupling mechanisms, and (3) if the modeling studies were properly validated. The last of the three was impossible to ascertain because historical records of a tsunami generated by massive flank failures of island stratovolcanoes is documented.

A review of present volcanic flank instabilities and flank collapses in the past (Pararas-Carayannis 2002) revealed massive slides that occurred in the Canary Islands (Day et al. 1999), Cape Verde (Day et al. 1999; Elsworth and Day 1999), the Hawaiian Islands (Moore et al. 1989, 1994a; Moore and Clague 1992; Lipman 1995), and other places. These events are documented in literature. However, the mechanisms of these prehistoric failures are not completely understood. Island stratovolcanoes appear to slide along their bases more frequently, and occasional locking of these slides may trigger large slope failures or earthquakes. Cumbre Vieja is composed of layers of pillow lavas interspersed with pyroclastic material (Pararas-Carayannis 2002).

The island of La Palma was formed by three statovolcanoes that rise several thousand feet above the sea floor. The volcano Taburiente lies to the north, and the central and southern parts of the island were formed by the volcanoes Cumbre Nueva and Cumbre Vieja. More than half of the subaerial volume of the island was removed by landslides and erosion during the past 1 million years (Pararas-Carayannis 2002). The Cumbre Nueva giant landslide, 48.8 mi^3, 7187 billion ft^3 (200 km^3), in volume, occurred approximately 560,000 years ago (Carracedo et al. 1999) that removed material in the west-central part of the island. The island is composed of two main rock layers separated at an elevation of approximately 1400 ft (427 m) above mean sea level (Pararas-Carayannis 2002). The thickness of pillow lavas ranges from 32.8 to 1148 ft (10 to 350 m). The upper layer consists of basaltic lavas and pyroclastic materials. Extensive erosion has taken place on the north part of the island, and to a lesser extent on Cumbre Vieja's flanks. Pararas-Carayannis (2002) concluded that the existic basaltic flows and dikes would significantly limit the volume of any future slope failure, and any such failure would either occur in steps or contained by ring dikes. The author did not find any evidence of a massive failure of the western flank of Cumbre Vieja along a deeper detachment surface. There is no extensive fault system along Cumbre Vieja's rift zone. The 13.1 ft (4 m) offset near the summit was suggested to be along a detachment fault by Ward and Day (2001). However, Pararas-Carayannis (2002) found no seismic or geologic data to support this suggestion; the offset could also have resulted from superficial gravitational settling or from collapse of a magmatic chamber.

Pararas-Carayannis (2002) evaluated the monolithic block movement of the Cumbre Vieja flank failure suggested by Ward and Day (2001). Flank collapses of island stratovolcanoes can be triggered by isostatic load adjustments, erosion, build-up of gaseous, hydrothermal, or magmatic pressure, violent eruptions, or collapse of magmatic chambers. An evaluation of the forces needed to cause the massive failure along a detachment surface suggested by Ward and Day (2001) was carried out by Pararas-Carayannis (2002). A force close to the base of the mass, or at least near the center of the mass, is needed to trigger a monolithic collapse due to exceedance of the shear strength along a slide plane. PararasCarayannis (2002) found that a force needed to move a 122.1 mi^3, or 17,969 billion ft^3 (500 km^3), monolithic block is unrealistic. There is also no evidence of significant magmatic-chamber collapse along the crest of Cumbre Vieja (PararasCarayannis 2002). The erosion on volcanic islands results in deposition of unconsolidated sediments. Large accumulations of these unconsolidated deposits can slide under gravity due to ground motion during earthquakes. On the La Palma Island, large amounts of sedimentary material, primarily gravel mixed with basaltic lava has accumulated on the western slope due to erosion of the Taburiente caldera, where a large surface landslide can occur due to a large earthquake, with the existing volcanic dikes providing some stabilization to these sediments (Pararas-Carayannis 2002). The author concluded that a large monolithic-slope failure of the dimensions suggested by Ward and Day (2001) is unlikely. A review of some other mechanisms also resulted in the same conclusion (Pararas-Carayannis 2002).

Pararas-Carayannis (2002) also evaluated the slide kinematics proposed and used by Ward and Day (2001) for their mega-tsunami simulation. The author found that the slide was assumed to be a monolithic rotation along a detachment fault rather than a turbulent flow of pyroclastic and pillow-lava material of large sizes. The monolithic-slide model also ignored the effect of cohesion among the particles within the mass that would resist movement and the effects of water turbulence behind the mass that would slow down the slide velocity. Rotational movement along a detachment fault should also be characterized by a reduction in slope with depth, which was not accounted for in the slide model of Ward and Day (2001). The maximum speed of the Grand Banks slide was estimated (Fine et al. 2005) at about 37.5-62.5 mi/hr (60-100 km/hr). It consisted of unconsolidated deposits rapidly moving downslope as a turbidity current. Ward and Day (2001) used a maximum slide speed of 328 ft/s (100 m/s), or 225 mi/hr (360 km/hr), far in excess of the maximum speed of the Grand Banks slide. Pararas-Carayannis (2002) argued that the mostly large- sized particles composed of pyroclastics and pillow lava of Cumbre Vieja cannot move as fast as turbidity currents.

Pararas-Carayannis (2002) stated that the tsunami generation and propagation model used by Ward and Day (2001) used several assumptions that resulted in overestimation of the tsunami wave heights. The initial waveform was estimated by Ward and Day (2001) based on a monolithic 122.1 la, or 17,969 billion ft^3 (500 km^3), block moving rapidly with a maximum velocity of 328 ft/s (100 m/s) for 37.5 mi (60 km) until it reached flat ocean bottom at a depth of 13,123 ft (4000 m). Pararas-Carayannis (2002) argued that incorrect assumptions leading to exaggerated source dimensions, slope instabilities, and slide speeds resulted in overestimation of initial tsunami-source parameters. The tsunami- propagation model used by Ward and Day (2001) treated the wave as a shallow-water wave, ignoring the amplitude attenuation away from the source (Pararas-Carayannis 2002). Also, as stated by Mader (2001), even the slide following the massive flank collapse proposed by Ward and Day (2001), with its ove maximum period of approximately 3 to 4 minutes, which will behave as an intermediate wave rather than a shallow water wave. Shallow-water waves undergo a geometric spreading, but not the significant wavelength-dependent dispersion (Pararas-Carayannis 2002) that characteristically affects shorter-period waves.

Wynn and Masson (2003) pointed out that the understanding of the initial stages of a landslide, which influence its tsunamigenic characteristics, is limited. Therefore, models of landslide-generated tsunamis remain highly dependent on poorly defined landslide parameters. The authors suggest that this understanding may be improved by studying the turbidite[(a)] deposits that are directly linked to known volcanic-island landslides.

Wynn and Masson (2003) obtained piston cores from two sites in the Agadir Basin, located about 187.5 mi (300 km) north of the Canary Islands, at a water depth of approximately 14,764 ft (4500 m). In the core sequences, the two youngest turbidites rich in volcanic material are correlated with the two most recent landslides on the Canary Islands. The correlations were made using dating of the turbidites, mineralogy of the turbidites, geochemical data, and analysis of seafloor morphology to define pathways of the turbidity currents. Both sequences showed layered deposits with layers of fine particles located higher in the sequences. The layer near the bottom, which contained the coarsest material, was the thickest. The two turbidites were separated from other turbidites by well-defined layers of pelagic or hemipelagic[(b)] sediment layers indicating separation of the deposition time of the turbidites of 1000 years or greater. Within each turbidite, no pelagic or hemipelagic sediment layers occur. Bioturbation[(c)] occurs in both turbidites, and the extent of bioturbation decreases downward within the turbidite. Based on these observed features of the turbidites, Wynn and MassOn (2003) concluded that the layers within the turbidites were not separated by substantially large time intervals, and, therefore, are all linked to the same turbidity-flow event. Wynn and Masson (2003) tentatively suggested that the layers within the turbidites linked to Canary Island landslides represent multiple stages of failure of a

---

(a) Sedimentary deposits formed by turbidity currents in deep water at the base of the continental slope and on the abyssal plain, first described by Bouma (1962).

(b) Pelagic sediments are of marine origin that accumulation in deep abyssal plains far from terrestrial sources of sediments; hemipelagic sediments consist of both terrestrial and *marine* sediments and occur closer to the continents.

(c) Bioturbation refers to the mixing of benthic flora and fauna with sediment particles.

single landslide event. The turbidite correlated to the El Golfo landslide on the northwest flank of the El Hierro island exhibited three layers; the turbidite correlated to the Icod landslide on the north flank of the Tenerife island showed nine layers. The authors interpreted the three layers of the first turbidite as indicative of a three-stage failure of the El Golfo landslide, and the nine layers of the second turbidite as indicative of a nine-stage failure of the Icod landslide. The authors also examined and discounted other possible explanations for layering of the turbidites, including flow reflection from basin margins or seamounts, flow surging or eddy formation, and transport through multiple channels. The authors also estimated that each layer would take at least two days to deposit based on an analysis of the grain sizes, the thickness of the layers, and the settling velocity of fine-grained sediment.

Wynn and Masson (2003) also attempted to assess the mechanism that created the individual turbidity currents that resulted in the layering of the (Masson et al. 2002), an argument could be made that the sediment was deposited by a single event.

As demonstrated by the summary of current research literature, the study of tsunamigenic landslides, although advancing rapidly, is not sufficiently mature to provide readily applicable engineering solutions. Each phase of a tsunami generated by a landslide should be studied with the help of experts in this area on a case-by-case basis, as required at a proposed nuclear power plant site.

2. Hierarchical Hazard Assessment Approach

2.1. Introduction

A hierarchical-assessment approach consists of a series of stepwise, progressively more refined analyses to evaluate the hazard resulting from a phenomena at a given nuclear power plant site. If the safety of a nuclear power plant can be demonstrated by a simple and bounding analysis, the resources and time required by more-refined assessment methods can be saved. However, the simpler analysis should meet the requirements of GDC 2 and all other applicable Nuclear Regulatory Commission regulations.

With respect to tsunamis, the hierarchical-hazard-assessment approach could be considered as a series of three steps:

1. Is the site region subject to tsunamis?

2. Is the plant site affected by tsunamis?.
3. What are the hazards posed to safety of the plant by tsunamis?

The first step above can also be regarded as a regional screening test. If the site region is not subject to tsunamis, no further analysis for tsunami hazards is required. Absence of credible tsunamigenic sources in the site region may result in a determination that the site region is not subject to tsunamis. However, this finding should be supported by region-specific evidence. If the answer to the first question is affirmative or undetermined— based on available information— an analysis of the tsunami hazard is required (i.e., the second step of the hierarchical-assessment approach should be performed).

The second step can be regarded as a site-screening test. This step determines whether safety system control important to safety of the plant are exposed to hazards from tsunamis. It may be possible to determine that, even though the general site region is subject to tsunamis, the plant itself is sited and designed in such a way that its safety is not affected. For example, if all SSC important to safety of the plant are located at an elevation above the maximum wave runup due to the PMT, more specific tsunami-flooding assessment may not be needed for the site.

The third step is the most refined assessment, in which site-specific analyses are carried out to determine hazards posed by the PMT to the SSC important to safety of the plant and to determine whether any protection is required. This step involves postulation of PMT source mechanisms, estimation of PMT source characteristics, initiation of the PMT wave, propagation of the PMT wave from the source toward the site, and estimation of tsunami hazards at the site.

The term tsunami typically relates to an oceanic tsunami generated by submarine seismic, landslide, or volcanic sources. However, tsunami- like waves can also be generated in inland water bodies by subaerial landslides and seismic causes [e.g., the 1811-1812 tsunami-like waves caused by the New Madrid earthquakes on the Mississippi River (Lockridge et al. 2002); the potential of a tsunami caused by earthquakes in Lake Tahoe (Ichinose et al. 2000); the tsunami in Spirit Lake during the May 18, 1980, eruption of Mount St. Helens (USGS 2007)]. The regional screening test should consider the likelihood of occurrence of a tsunami or a tsunami-like wave in a water body near the nuclear power plant site.

2.2. Regional Screening Test

A regional survey and assessment of tsunamigenic sources should be performed to determine the potential that a tsunami may pose a hazard to the site. Significant hazards to the site may arise both from near-field and far-field tsunamis. The regional survey and assessment should include all potential far-field sources and mechanisms that generate tsunamis.

Nuclear power plant sites in the United States can generally be classified into two categories: coastal sites (located on or near a coastline) and inland sites (located at significant distance from the coastline). Coastal sites should consider hazards from oceanic tsunamis; inland sites should consider the possibility of a tsunami-like wave in water bodies in the region.

The regional screening test involves the following steps:

1. a comprehensive search for historical and paleotsunami records in the region
2. in the absence of historical tsunami records, a comprehensive search for potential tsunamigenic sources and the potential for tsunami generation in nearby water bodies.

Sources of historical and paleotsunami data are described in Chapter 4, Databases and Data Collection, of this report. A comprehensive search of these national and international (where needed) data repositories should be carried out to list all historical tsunamis that occurred in the region. When available, wave height, inundation extent, runup, and drawdown associated with these events should be described. Paleotsunami data and inferences drawn by experts from these data should also be included in the report because these data-sets extend the historical record and may include events more severe than those actually recorded.

The fact that no historical tsunami records can be found for a region does not necessarily result in a conclusion that the region is free of tsunami hazards. A comprehensive search for potential tsunamigenic sources that may create a tsunami in regional water bodies should be carried out. An example of such a study is described by Gonzalez et al. (2003).

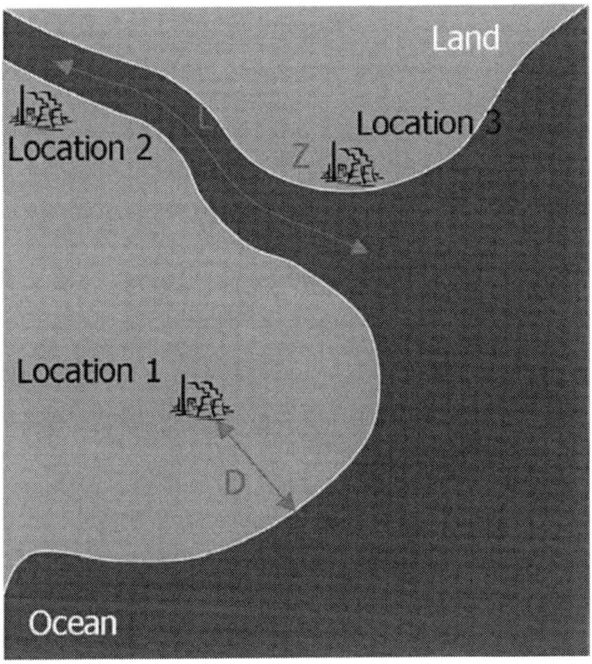

Figure 2-1. The DLL screening rule.

2.3. Site Screening Test

One possible way to answer the question in the second step of the hierarchical-assessment approach described above is to use a site-screening test to compare the location of the plant site with the area affected by tsunamis in the region. Figure 2-1 shows three possible locations of a plant located near a coast. Location 1 could be safe if the horizontal extent of inundation caused by the probable maximum tsunami was less than the horizontal distance (D). The horizontal extent of inundation varies depending on the characteristics of the tsunami waves, shoreline, and beach geometry and topography, and the effective obstruction to the inland flow of tsunami waves on the shore. According to the NGDC historical tsunami database (NGDC 2007), the maximum horizontal extent of inundation is listed as 3.4 mi (5.5 km) for the December 26, 2004, Indian Ocean tsunami on the island of Sumatra, Indonesia. Anecdotal evidence indicates that the maximum extent of horizontal inundation on the island may have reached 5.0 mi (8.0 km).

Location 2, located on a river near an estuary, could be safe if the backwater effects or bores induced by the PMT affected areas in extent less than the longitudinal distance (L) measured along the river. Bores can be induced by a tsunami (Koch and Chanson 2005) and were observed in Hawaii in 1946, in Japan in 1983 and 2003, and in Thailand, Malaysia, and Sri Lanka in 2004. Bores propagating upstream from the mouth of a river may be caused by tsunamis under favorable hydraulic conditions and can travel upstream several tens of miles (km) from the estuary (Chanson 2005). Bores are similar to traveling hydraulic jumps that move upstream. The downstream part of the bore is characterized by subcritical flow with greater depth of flow and the upstream part is supercritical. The river should be able to support supercritical flow to form the bore.

Location 3, a site located on the coast, could be safe if the runup caused by the PMT did not reach its grade elevation (Z) and all SSC important to safety of the plant are located at or above this grade elevation. In the NGDC historical tsunami database, the runup is defined as the maximum ground elevation that the tsunami waves reach inland from the shore. The maximum runup reported in the NGDC historical tsunami runup database, 1720 ft (524 m), is associated with the massive 1958 Lituya Bay subaerial landslide [more than 1 billion cubic ft (30 million cubic m) in volume] and tsunami (Miller 1960). The database contains runup heights of 820 ft (250 m) and 738 ft (225 m) above the surface of the Spirit Lake on the west and east arms of the lake caused by the collapse of the upper 1500 ft (460 m) of the dome of Mount St. Helens and subsequent gigantic debris fall into the lake (Waitt and Pierson 1994). The Lituya Bay slide occurred on one slope of a narrow bay, causing the unusually high runup on the opposite slope. The enormous debris fall from collapse of the Mount St. Helens dome was the cause of the unusually high runups in Spirit Lake. Oceanic tsunamis have caused runups as high as 328 ft (100 m) in Indonesia (Banda Sea earthquake and tsunami of 1674). The maximum runup caused by the 2004 Sumatra tsunami was 167 ft (50.9 m), observed on the Sumatra island.

The preceding discussion clarifies that the severity of tsunami hazards is strongly affected by local characteristics. Therefore, accurate estimation of hazards posed by tsunamis to a nuclear power plant site requires site-specific analyses. However, it is also clear that unusually high tsunami runups have only been recorded for unusually rare events (e.g., Lituya Bay and Spirit Lake tsunamis). If these extremely rare events are excluded, even the largest historical tsunamis do not exceed runup of approximately 328 ft (100 m) and do not inundate areas more than a few miles (km) inland from the shore. Also, tsunami- induced bores do not travel more than a few tens of miles (km) upstream from the mouth of a river. Therefore, it may be possible to screen sites based on a combination of historical record, site-specific geophysical and topographical data, and sound engineering judgment. Detailed site-specific tsunami hazards assessments should be completed for sites that would be affected by tsunamis and for sites for which a conclusive determination cannot be made that they are free of tsunami hazards.

A situation similar to that shown in Figure 2-1 may also exist for inland sites located near a body of water in which a tsunami may be caused by seismic, landslide, or volcanic sources. The DLZ rule may still be applied to the inland sites to determine if the site is safe from the hazards posed by tsunamis.

2.4. Detailed Tsunami Hazard Assessment

If the site-screening test does not establish the safety of the nuclear power plant site from hazards posed by tsunamis, a detailed assessment of these hazards should be undertaken to ensure that the plant design bases adequately account for these hazards.

A companion to this report is the National Oceanic and Atmospheric Administration Pacific Marine Environmental Laboratory (PMEL) Technical Memorandum OAR PMEL-136, titled "Scientific and Technical Issues in Tsunami Hazard Assessment of Nuclear Power Plant Sites" (González et al. 2007), which describes the state-of-the-art in tsunami-hazard assessment, including data sources, paleotsunami identification approaches, and tsunami propagation and inundation simulation. Gonzalez et al. (2007) should be used as a reference for technical bases for a detailed tsunami-hazard assessment.

2.5. Site Investigation

For a detailed tsunami-hazard assessment at a site of interest, a thorough investigation is needed to establish the history of tsunamis at a given nuclear power plant site and to collect site-specific data required for estimation of the hazard metrics. These investigations should lead to the determination of historically recorded tsunamis and compilation of known pre-historical records. An analysis of all tsunami data should be carried out to determine the potential mechanisms that may generate a tsunami capable of affecting the site. The analysis should also specify tsunami characteristics and the corresponding damage at the site. The frequency and severity of tsunamis at the site from historical and prehistorical data should be estimated.

Because both near- and far-field tsunamis require investigation, the compilation of historical observations and prehistorical data may need to access international databases as well as those archived and maintained in the United States.

2.5.1. Historical Tsunami Records

The first step in site investigation is establishing the history of tsunamis at and near the proposed site. All available historical data related to tsunamis should be collected. Where available, these data should include information regarding

1. date and time of occurrence
2. source mechanism
3. source location
4. source parameters
5. wave height on bottom pressure gauge and tide records
6. tidal records or ambient water levels at the time of tsunami arrival
7. tsunami runup
8. tsunami inundation (horizontal distance the waves reached from the shore)
9. damages caused.

The NGDC historical tsunami database is the most comprehensive compilation of tsunami events. The database should be searched for historical events in the vicinity of the site. The NGDC tsunami database also provides metadata regarding the tsunami events and references to where these events were first reported or analyzed. The meta- data often hold much additional information about the nature of tsunami observation, the source mechanism, and damages caused by the event. A careful review of the metadata and references cited should also be compiled for the site of interest.

Some of the tsunami-like events are also listed as seiches in the NGDC database. These events can be selected using the "type of measurement" search criteria set to "seiche." Tsunami events caused by volcanic activity are cross-linked to the NGDC volcanoes database, which can provide additional information regarding the tsunamigenic volcanic activity.

Seismic activity can trigger landslides, and tsunamis can be generated by both these source mechanisms and may arrive at a site within a short period of time. The NGDC database should be queried for such events, and all references listed in the metadata should be searched to determine the sequence of events. These records, even if anecdotal, can provide significant information regarding concurrence of source mechanisms and consequent enhancement of tsunami impacts at the site.

Based on the collected historical data, tsunami activity at and near the site should be thoroughly described. This description should list all observed source mechanisms, the ranges of source parameters and characteristics, the ranges of tsunami runup and inundation, and the extent of damage suffered.

2.5.2. Paleotsunami Evidence

Because the recorded history of tsunamis in the United States is relatively short, there are insufficient data related to, indeed even complete absence of, tsunami activity, depending on the location of the site. The problem is exacerbated because large and destructive tsunamis are relatively infrequent events. One way to supplement the relatively short observed record is the evidence provided by paleotsunami data (see Chapter 3 of Gonzalez et al. 2007).

At this time, no central database of paleotsunami information for use in the United States exists. One such effort is described by Peters et al. (2003) that focuses on the Pacific Northwest coast and the tsunamis originating in the Cascadia subduction zone.

It may be necessary to carry out a site-specific data collection to seek geologic evidence of tsunamis near the proposed nuclear power plant site. In general, sites within approximately 62.5 mi (100 km) of the coast should be searched for tsunami and boulder deposits in locations likely to preserve tsunami deposits, such as coastal marshes and shallow lakes. However, expert assistance may be needed for this search and subsequent interpretation of the candidate deposits (González et al. 2007).

2.5.3. Regional Tsunami Assessments

A regional tsunami-hazard assessment, previously carried out by other responsible agencies, may be available for areas that are frequently subject to tsunamis. A thorough search should be carried out to identify and evaluate the applicability of such assessments to the proposed nuclear power plant site. Most tsunami-hazard-assessment studies are carried out from the point of view of flooding hazard from inundation and potential for loss of life and damage to property and infrastructure. The tsunami hazards assessment for a nuclear power plant site needs to address several aspects of the hazard from tsunamis (see Chapters 3 and 5 of this report). The existing regional tsunami- hazard assessment may be used as a starting point for a more focused site-specific assessment.

González et al. (2006) describe a pilot study of tsunami hazards to update Federal Emergency Management Agency flood hazards maps for Seaside, Oregon (see Appendix D of Gonzalez et al. 2007).

The González et al. (2006) study was carried out for the area near the towns of Seaside and Gearhart on the Oregon coastline. This area was chosen for the study because it is similar to many coastal communities and because there was a strong local interest in such a study. The study developed two inundation maps, one for the 100-year-return-period tsunami and the other for the 500-yearreturn-period tsunami. The components of the study were:

1. source specification: based on literature review and consultation with tsunami experts, a database of quantitative probabilistic models of local and far-field earthquake tsunami sources in the Cascadia, Alaska-Aleutian, and Peru-Chile subduction zones was developed
2. data acquisition: a paleotsunami deposit mapping and interpretation study was carried out, along with acquisition of historical records and eyewitness accounts
3. model development: a new, high-resolution digital elevation model (DEM) based on the latest available topography, bathymetry, and tidal information was developed. A site- specific tsunami inundation model was developed and tested with available tsunami observations, including paleotsunami data, historical records, and eyewitness reports
4. probabilistic tsunami hazard estimation
5. study-specific database development
6. analysis and interpretation of tsunami impacts.

In the context of tsunami-hazard assessment at proposed nuclear power plants for determination of safety, a similar procedure may need to be adopted, with the notable exception of the use of a PMT in the place of a chosen return period to specify the design tsunami and the corresponding probabilistic hazard estimation. Also, the González et al. (2006) study only used earthquake sources for tsunami generation. The tsunami- hazard assessment at a proposed nuclear power plant site will also need to include landslide and volcanic sources.

As illustrated above, regional tsunami-hazard assessment studies should be carefully evaluated to ascertain their applicability to tsunami-hazard assessment at a proposed nuclear power plant site. These studies can serve as a starting point, and may provide valuable sources of data focused on the needs of the safety determination at the site.

2.5.4. Site-Specific Tsunami Mechanisms

For a proposed nuclear power plant site, a thorough search should be undertaken to catalog sources of all known (observed historical tsunami event or thoroughly analyzed and agreed-on paleotsunami record) and proposed (based on paleotsunami evidence) tsunami events at and near the site. These sources should include both near- and far-field sources. All three tsunamigenic source mechanisms should be investigated.

For each source mechanism, a list of candidate tsunamigenic sources should be created. Information on source parameters should be collected using literature surveys and contacting agencies and organizations with relevant expertise. Experts in the relevant fields may need to be consulted for estimation of parameters when no published values exist. Sources known to have generated tsunamis in the past should already be available from the compilation of historical tsunami records recommended by Section 1.4.1. The observed tsunami data and known or estimated parameters of these known sources should be used to validate tsunami simulation models.

In addition to compiling the parameters (known or estimated), particular attention should be given to the orientation of the source with respect to the proposed nuclear power plant site. The orientation of the source, particularly landslides, can strongly affect the directivity of the tsunami waves. Therefore, even a relatively moderately strong tsunamigenic source may result in a greater hazard to the site than a stronger source that is not optimally oriented to the site of interest. Similarly, particular attention should be paid to sources near the site because any generated tsunami waves from these sources would have little attenuation compared to waves that travel to the site from a great distance. Source parameters that are important for tsunami generation and subsequent hazard estimation are described in Chapter 5 of this report.

2.5.5. Site-Specific Data

Site-specific geospatial and geophysical data are required for accurate simulation of the near-shore dynamics of tsunami waves and the estimation of tsunami-hazard metrics. A selection of these hazard metrics are described in Section 5.4 of this report. Site-specific geospatial data that are required for accurate simulation of the tsunami- wave dynamics primarily consist of fine-scale maps of near-shore topography and bathymetry. Data sources for topography and bathymetry are described in Section 4.2 of this report. Site-specific geophysical data are related to specification of tsunamigenic source characteristics and to estimation of tsunami-hazard metrics. Tsunamigenic source characteristics are described in Section 5.3 of this report.

3. Effects of Tsunami at a Nuclear Power Plant Site

3.1. Introduction

Tsunamis can result in a severe hazard to safety- related cooling-water systems as well as other SSC important to safety of a nuclear power plant. The primary effect of the tsunami waves on a plant site is flooding (directly from tsunami waves, or from backwater effects or tidal bores caused upstream from the mouth of a river, depending on the location of the site) and loss of cooling water (due to dry intakes during drawdown caused be receding tsunami waves). However, there are also several other effects, mainly from hydrodynamic forces that can cause severe damage to structures and the foundations of these structures. If any of these structures are safety-related, they should be designed to withstand these effects or be protected adequately from these effects to ensure the safety of the nuclear power plant.

3.2. Flooding Due to Runup

The most obvious hazard from a tsunami is flooding. As the tsunami waves approach the shoreline, they increase in amplitude and reduce in wavelength. The waves, depending on local bathymetry and topography, can inundate significantly large areas inland from the shoreline. The maximum ground elevation that the inundating waves reach is called the runup.

The protection from flooding effects of tsunami waves can be provided in a manner similar to other flooding mechanisms: either the SSC important to safety can be located above the maximum runup due to the PMT or adequate flooding protection for the SSC important to safety can be provided to ensure that function is not compromised. Belts of trees and mangroves provide barriers to tsunami by effectively increasing frictional resistance to wave runup.

The feasibility and effectiveness of such measures should be investigated.

3.3. Dry Intakes During Drawdown

During recession of the tsunami waves (alternatively, approach of the depression wave), water level at the shoreline is lowered. The amount of lowering depends on local bathymetry. The areal extent of recession can be significantly large. Nuclear power plants that depend on an intake or intakes that are located offshore for their safety- related cooling water needs should ensure that the maximum extent of recession and the accompanying lowering of the water level near the intake location do not result in dry intakes.

Protection from receding tsunami waves can be provided in a manner similar to other low-water mechanisms: either the safety-related intakes can be located sufficiently away from the shoreline and in deeper waters that the recession and the accompanying lowering of the water level does not result in dry intakes, or alternative sources of safety-related water supply can be made available that are independent of the water body that is affected by the receding tsunami waves.

3.4. Scouring

The behavior of the tsunami waves in shallow waters near the shore is very complex. Tonkin et al. (2003) carried out experiments to investigate the scouring effects of tsunamis around cylindrical structures. They found that the scouring at the front (facing the oncoming waves) of the cylinder could be explained by a standard shear-stress model. However, they also observed rapid scouring in the sand substrate at the back of the cylinder at the end of the tsunami drawdown, when flow velocity was decreasing rapidly. This rapid scour created the largest and deepest scour holes. The investigators proposed that the observed scour at the back of the cylinder may be explained by considering the effects of pore pressures in the substrate during the tsunami. The rapidly decreasing water level at the end of the tsunami drawdown may create large-pore pressure gradients, resulting in a buoyant force on the substrate. When these buoyant forces are large enough to decrease the effective normal load substantially and result in loss of shear strength, even low-flow velocities of the tsunami currents may result in rapid scouring. In extreme cases, a complete loss of shear strength may occur, leading to liquefaction of the substrate.

A nuclear power plant that locates any safety-related structures, particularly a safety-related intake structure, where tsunami currents may potentially result in scouring and the resulting damage to the foundations of such structures, should ensure that the structures are adequately designed to resist the scouring forces of tsunamis. An obvious alternative is to provide a source of safety-related cooling water that is independent of the water body experiencing the tsunami and, therefore, is not exposed to the scouring effects of tsunamis.

3.5. Deposition

Tsunami currents near the shoreline are highly turbulent, capable of carrying debris and sediment. As the waves recede, the sediment and debris can be deposited at and near the shoreline. For a nuclear power plant, all SSC important to safety should be located and designed such that they are not affected by the deposition of debris and sediment from tsunami waves and currents.

3.6. Hydrostatic and Hydrodynamic Forces

Hydrostatic force acts laterally on structures. It can result from standing water resulting from tsunami inundation or from interaction of tsunami water moving slowly and encountering the structures.

Hydrostatic forces will also be experienced by intake structures that are located offshore. The intake structure may experience a cyclic hydrostatic loading as tsunami waves pass over it. The hydrostatic pressure during the passage of a crest will be higher and lower during passage of a trough due to normal still-water level (Yeh et al. 2005).

Hydrodynamic forces result from rapidly moving water and its interaction with structures. On-shore structures may experience impacts on the sides facing the oncoming wave, drag forces on the sides, and suction on the downstream end of the structure (Yeh et al. 2005).

Yeh et al. (2005) provide a set of generalized expressions for wave and flood loads on structures. These include expressions for hydrostatic force, buoyant force, hydrodynamic force, surge force, impact force, and breaking-wave forces. Loading combinations are also described. Evaluation of tsunami loadings on a reinforced concrete building is also provided (see the appendix in Yeh et al. 2005).

3.7. Debris and Projectiles

Tsunamis are capable of dislodging and transporting a wide range of debris from sediment to large boulders (Mastronuzzi and Sanso 2000) to other water-borne projectiles (automobiles, trees, boats, etc.). Debris and projectiles can compound the impacts on structures. Yeh et al. (2005) describe the behavior of water-borne projectiles. Projectiles are considered to impact the structures at the same elevation as the water-surface level. The uncertainty in the estimation of the duration of the impact is thought to be the most likely cause of errors in the estimation of the impact forces. The duration of impact forces, however, are affected primarily by the natural frequency of the structures (Chopra 1995).

Adequate design criteria should be employed for SSC important to safety exposed to impacts from water-borne debris and projectiles. An alternative is to locate SSCs so that they will not be exposed to water-borne debris and projectiles.

3.8. Tidal Bores

Tidal bores are similar to traveling hydraulic jumps that move upstream from the mouth of the river. They are generally caused by a rapid and large change in the downstream water-surface elevation (Chanson 2005; Koch and Chanson 2005).

Tidal bores can be induced by a tsunami (Koch and Chanson 2005), and were observed in Hawaii in 1946, in Japan in 1983 and 2003, and in Thailand, Malaysia, and Sri Lanka in 2004. Tidal bores propagating upstream from the mouth of a river may be caused by tsunamis under favorable hydraulic conditions and can travel upstream several tens of miles (km) from the estuary (Chanson 2005). The downstream part of the bore is characterized by subcritical flow with greater depth of flow, and the upstream part is supercritical. The river should be able to support supercritical flow to form the bore.

The effect of a tidal bore propagating upstream to a nuclear power plant site is similar to a flood wave propagating downstream. This may result in flooding of the site depending on the height of the bore. Bores dissipate as they travel upstream. If the site is sufficiently far away from the estuary, the bore may not flood the site. Siting and protection criteria for tidal bore are similar to those for other flooding mechanisms.

4. Databases and Data Collection

4.1. Introduction

In the United States, tsunami records are archived by the NGDC, which is one of the three environmental data centers within the National Environmental Satellite, Data, and Information Service. The NGDC also operated World Data Centers for Marine Geology and Geophysics and Solid Earth Geophysics. Together, WDC and NGDC acquire, process, and distribute global marine and terrestrial data. They also have a role in post-event data collection of tsunami sources and effects that supports modeling, engineering, planning, and educational goals.

The primary access to the NGDC data archive is via the internet at http://vvww.ngdc.noaa.gov. This website archives data on natural hazards (earthquakes, tsunamis, and volcanoes), marine geology and geophysics (including ocean drilling data, well logs, grain size data, and sediment thickness), and bathymetry and topography data [including combined bathymetry and topography datasets, multibeam data, National Ocean Service (NOS) hydrographic surveys, and global relief data (ETOPO2, ETOPO5, etc.)].

4.2. Topography and Bathymetry

One of the most important datasets for use in tsunami modeling is bathymetry and topography. Tsunami waves travel at different speeds in waters of different depths, and, therefore, the wave front is modified as the waves propagate in a water body with varying depth. In deep waters where the tsunami waves have long wavelengths, the grid points needed to resolve the wavelength in numerical models can be spaced far apart; therefore, a relatively low resolution relief data may be used (e.g., the 2 and 5 arc-second digital global relief datasets ETOPO2 and ETOPO5) without compromising the accuracy of model predictions. However, in shallower waters, the wavelength of the tsunami waves shorten. Thus, accurate and high-resolution relief data are needed (e.g., high- resolution NOS surveys) to resolve the wave and to ensure accuracy of model predictions. Topography data may be even more important because inundation models typically have the most uncertainty in their predictions.

4.2.1. Topography Data

Digital topography data have traditionally been created, verified, and distributed by the USGS (http://edc.usgs.gov/geodata/). The USGS topography data are available at 1:24,000 scale for 7 5-min quadrangles. The digital data are also available at 32.8-ft (10-m) horizontal resolution with a nominal vertical accuracy of 3.3 ft (1 m). Traditionally, USGS digital topographic data do not extend below the surface of the water into water bodies. Areas covered by the water bodies are typically shown at some nominal constant elevation above the mean sea level.

Currently, the best quality digital topographic data can be created using airborne light detection and ranging (LiDAR) technology. LiDAR is an active imaging system similar to a radar. It transmits laser pulses to a target and measures the time for the signal to be reflected back to the sensor. The LiDAR-mapping flights are usually flown at altitudes of 1000-6500 ft (300-2000 m) and use near-infrared light (1.045-1.065 1.tm). Areas approximately 70 percent of the altitude in width can be covered. Horizontal resolutions as small as 2.5 ft (0.75 m) with a horizontal accuracy of approximately 3.3 ft (1 m) can be obtained. The vertical accuracy can be approximately 0.5 ft (0.15 m). However, in many cases, dedicated LiDAR mapping may be required to obtain such data, and costs can be significant.

The NOAA Coastal Services Center (CSC) acquires high-resolution topographic data using remote sensing under the topographic-change mapping project since 2001 (see the website at http://www.csc.noaa.gov/crs/tcm/). The topographic-change mapping project followed the Airborne LiDAR Assessment of Coastal Erosion (ALACE) project, a partnership between the NOAA CSC, the National Aeronautics and Space Administration Observational Sciences Branch, and the USGS Center for Coastal Geology, that collected LiDAR data along the U.S. coastline from 1996 until 2000 using the NASA Airborne Topographic Mapper sensor. The NOAA CSC distributes digital elevation data derived from two sensing technologies: LiDAR and IfSAR or InSAR, the Interferometric Synthetic Aperture Radar, which uses radar pulses. The data can be found at the CSC website (http://maps.cscmoaa.gov/TCM/).

4.2.2. Bathymetry Data

Digital global relief data are available from the NGDC at two resolutions: 2 arc-seconds (ETOPO2v2 dataset published in June 2006) and 5 arc-seconds (ETOPO5). The dataset covers the whole globe from -90° to +90° in latitude and - 180° to +180° in longitude. Both the cell- centered version (where the cell boundaries are lines of even minutes of latitude and longitude, centered on intersections of lines of odd minutes of latitude and longitude; a grid of 5400 rows and 10,800 columns) and the grid-centered version (where the cell boundaries are lines of odd minutes of latitude and longitude centered on intersections of lines of even minutes of latitude and longitude; a grid of 5401 rows and 10,801 columns) are available. Subsets of these datasets may be used for tsunami propagation modeling for far-field tsunamis.

The NOS hydrographic database (NOSHDB) is maintained by the NGDC and is located at the its website (*http://www.ngdc.noaa.gov /mgg/bathymetry/hydr*o.html). The NOSHDB provides extensive coverage of coastal waters and exclusive economic zones of the United States and its territories. Initially, the database was created by digitizing the sheets of hydrographic surveys completed between 1851 and 1965. Since 1965, NOS survey vessels have acquired data digitally. The database is available for interactive search and download on the internet at *http://map.ngdc. noaa.gov/website/mga/nos* hydro/ viewer.htm. The database may also be obtained as a stand-alone set of DVD-ROMs or CD-ROMs with included search software. Data for the near- shore inundation modeling domain may be downloaded for creation of a computational grid for a particular site. Depending on the capabilities of the tsunami-simulation software, NOSHDB data may be used to create nested grids within the ETOPO2v2 grid for propagation and inundation simulations.

The NGDC is also developing high-resolution combined bathymetry and topography DEM for applications of the Method of Splitting Tsunamis (MOST) model (*http://www.ngdc.noaa.gov/mgg/inundation/tsuna* mi/general.html). These DEMs are being developed using the best digital available data from federal and state agencies, academic institutions, and private companies.

Various sources of data are merged and supplemented when required by additional hand- digitized data. A set of quality-control checks is carried out for consistency of the final DEM to remove artifacts. Several DEM grids on the East Coast (Virginia Beach, Virginia; Savannah, Georgia; Myrtle Beach, South Carolina; Cape Hatteras, North Carolina; San Juan and Mayaguez, Puerto Rico), the coast of Alaska (Dutch Harbor and Sand Point), the Gulf Coast (Panama City, Florida), and the West Coast (Port San Luis, California) are already completed. The cell size of these DEMs is 1/3 arc-second [approximately 32.8 ft (10 m)].

Multibeam bathymetry data are available from the NGDC, which is the national archive in the United States (http ://www. ngdc. noaa. gov/mgg/bathymetry/multi bearn.html). The NGDC database contains over 1.6 million nautical miles (3 million km) of ship trackline data and hydrographic multibeam survey data. Bathymetric grids, three-dimensional images, and data can be downloaded from the NGDC. Several other sources of multibeam data also exist: the Canadian marine multibeam bathymetric data (http://gdr.nrcan.gc.ca/multibath/index_e.php), the Woods Hole Oceanographic Institution multibeam bathymetry data (http://mbdata.whoi.edu/mbdata.html*)*, and the Marine Geoscience Data System (http://www.marine-geo.org*/)*.

Bathymetry data collected using LiDAR and IfSAR by the NOAA CSC are available from the CSC (http://maps.csc.noaa.gov/TCM/) archived from the ALACE project and the continuing topographic-change mapping project. An evaluation of various publicly available global bathymetry datasets is presented by Marks and Smith (2006).

4.3. Tides and Sea-Level Anomalies

The Center for Operational Oceanographic Products and Services (CO-OPS) located within NOAA's NOS collects, archives, analyzes, and distributed a variety of oceanic and coastal data, including historical water-level data via the National Water Level Program (NWLP). The NWLP consists of a network of long- and short- term water-level stations. There are 175 long-term stations, 261 coastal stations, and 55 Great Lakes stations for which tide data are available at http://tidesandcurrents.noaa.gov/station_retrieve.shtml?type=Tide+Data. The database also contains historical tide data from stations located all over the world. Data can be retrieved for more than 2700 historical stations.

4.4. Tsunami Wave Heights., Runup, and Drawdown

The NGDC tsunami-source-event database is located on the Internet at the website *http://www.ngdc.noaa.gov/hazard/tsu.shtml.* The tsunami-source-event database is global in extent and contains information on tsunamis, including source description and location, date and time of occurrence, event magnitude, water height, and corresponding damages. The database can be searched online for source type, source location, time periods of interest, runup location, and effects of the tsunami. More than 2300 events are listed in the tsunami-source-event database.

The NGDC tsunami-runup database is located on the Internet at the website http://www.ngdc.noaa.gov/hazard/tsu_db.shtml. Similar to the tsunami-source-event database, the tsunami-runup database is global in extent. It contains each runup reported for all tsunamis in the tsunami-source database as a separate entry. For each runup, date and time, country, location, maximum water height, inundation distance, tsunami source parameters, and effects of tsunami runup are stored. More than 9200 runup events are listed in the database.

The databases allow several search parameters that can be effectively used to search for historical tsunami information for a location of interest. Metadata stored along with each record also includes a short description and lists references for each event and runup. Drawdown is not reported in the database. Typically, drawdown observations are difficult to obtain from post-event surveys and are generally inferred from eyewitness accounts or, in rare circumstances, from remote- sensing images.

4.5. Near-Shore Currents

Near-shore currents can be classified as tidal and non-tidal currents. Tidal currents are periodic and follow the tide cycle. Non-tidal currents or ocean currents include permanent currents in oceanic circulatory systems. Tidal current data are measured at several locations on the U.S. coastlines by NOS using Physical Oceanographic Real-Time System (PORTS), available from NOAA (http://tidesandcurrents.noaa.gov/station_retrieve.shtml?type=Current+Data).

Ocean currents are organized flows that persist over a time period in a geographical location in the ocean. These currents transport water, heat, chemicals, and organisms from one part of the ocean to another. Wind is one of the primary forces that drive these currents. Wind creates a surface stress, causing water particles to move. Over large distances, the flow is affected by the Coriolis force. Another cause of oceanic currents is difference in water density, which depends on temperature, salinity, and the pressure of surrounding water. The density differences cause what is known as thermohaline flow in the oceans. Some data related to open ocean currents are archived at the University of Miami Rosenstiel School of Marine and Atmospheric Science (http://oceancurrents.rsmas.miami.edu/).

4.6. Seismic Data

The NGDC earthquake database is located on the Internet at the website http://www.ngdc.noaa.gov/hazard/earthqk.shtml. It contains data on more than 5000 destructive earthquakes since 2150 B.C. in a catalog of significant earthquakes. A significant earthquake is defined as an earthquake that resulted in moderate damage ($1 million or more), a magnitude of 7.5 or more, 10 deaths, a modified Mercalli intensity[(a)] of X, or a tsunami.

[(a)] The modified Mercalli intensity is given in Roman numerals ranging from Ito XII. The values on the intensity scale are based on a subjective description of the effects of the earthquake:

I. Not felt except by a very few under especially favorable circumstances.

II. Felt only by a few persons at rest, especially on upper floors of buildings. Delicately suspended objects may swing.

III. Felt quite noticeably indoors, especially on upper floors of buildings, but many people do not recognize it as an earthquake. Standing motor cars may rock slightly. Vibration like passing truck. Duration estimated.

IV. During the day felt indoors by many, outdoors by The NGDC earthquake-intensity database contains information on more than 23,000 U.S. few. At night some awakened. Dishes, windows, and doors disturbed; walls make creaking sound. Sensation like heavy truck striking building. Standing motorcars rock noticeably.

V. Felt by nearly everyone; many awakened. Some dishes, windows, etc., broken; a few instances of cracked plaster; unstable objects overturned. Disturbance of trees, poles, and other tall objects sometimes noticed. Pendulum clocks may stop.

VI. Felt by all; many frightened and run outdoors. Some heavy furniture moved; a few instances of fallen plaster or damaged chimneys. Damage slight.

VII. Everybody runs outdoors. Damage negligible in buildings of good design and construction slight to moderate in well built ordinary structures; considerable in poorly built or badly designed structures. Some chimneys broken. Noticed by persons driving motor cars.

VIII. Damage slight in specially designed structures; considerable in ordinary substantial buildings, with partial collapse; great in poorly built structures. Panel walls thrown out of frame structures. Fall of chimneys, factory stacks, columns, monuments, walls. Heavy furniture overturned. Sand and mud ejected in small amounts. Changes in well water. Persons driving motor cars disturbed.

The NGDC earthquake-intensity database contains information on more than 23,000 U.S. earthquakes, including their locations, magnitudes, focal depths, and reported damages. The earthquakes listed in this database span the period 1638 to 1995.

The NGDC also distributes a global CD-ROM, database of over 4 million earthquakes from 2100 B.C. to 1995 A.D. Online databases can be interactively searched as required for a particular application.

Global earthquake data are also archived at the USGS National Earthquake Information Center (*http://earthquake.usgs.gov/regional /neic/index.ph*p). The database contains information on earthquakes from 2000 B.C. through the current week because it is updated dynamically. The database can be searched interactively on the website as required for a given site or region.

4.7. Geophysical Data

In addition to tsunami and earthquake databases, the NGDC also archives a volcano database (http ://www.n gdc. noaa. gov/hazard/volcano.shtml). The significant-volcanic-eruptions database contains information on more than 400 eruptions dating from 1750 B.C. The volcano-location database contains information on more than 1500 volcanoes. The tsunami database is cross-linked with the volcano database, indicating which observed tsunamis have a volcanic-source mechanism. The volcano database contains the volcanic-explosivity index, which is one of the source parameters for a tsunamigenic volcanic event.

Offshore tsunami waves are measured primarily by bottom pressure gauges that record data in real- time. In the 1980s, NOAA PMEL developed deep-ocean tsunameters for early detection and measurement of tsunami waves and application of this information to a warning system. This project

---

IX. Damage considerable in specially designed structures; well-designed frame structures thrown out of plumb; great in substantial buildings, with partial collapse. Buildings shifted off foundations. Ground cracked conspicuously. Underground pipes broken.

X. Some well-built wooden structures destroyed; most masonry and frame structures destroyed with foundations; ground badly cracked. Rails bent. Landslides considerable from river banks and steep slopes. Shifted sand and mud. Water splashed over banks.

XI. Few, if any (masonry), structures remain standing. Bridges destroyed. Broad fissures in ground. Underground pipelines completely out of service. Earth slumps and land slips in soft ground. Rails bent greatly.

XII. Damage total. Waves seen on ground surfaces. Lines of sight and level distorted. Objects thrown upward into the air.

was called Deep-ocean Assessment and Reporting of Tsunamis (DART; see http ://nctr .pmel. noaa. gov/Dart/index. html). A DART system consists of a bottom pressure recording (BPR) system connected to a surface buoy for real-time reporting of recorded data. The system can detect tsunami waves as small as 0.4 in. (1 cm). In December 2006, there were 25 fully functioning DART locations, 20 in the Pacific Ocean (including five along the U.S. West Coast and seven along the southern Alaska coast and Aleutian Islands), four in the Atlantic, and one in the Gulf of Mexico. There are an additional 14 locations planned for future DART deployment.

The edited BPR data from the DART system and historical BPRs are available from the NGDC for the period 1986 to 2004. Real-time DART data are available from the NOAA National Data Buoy Center (*http://www.ndbc.noaa.gov/dart.shtml).*

Currently, no databases (global or regional) are available that catalog submarine landslides and their characteristics. As described in Section 2.4 of this report, extensive site-specific geological, seismic, and geotechnical data, some using advanced imaging techniques, may be needed to completely characterize submarine landslides. The interpretation of these datasets is also an advancing area of research. As more accurate and higher-resolution bathymetry and submarine geophysical data are compiled, offshore historical landslides and zones of potential landslides can be more readily identified (see Mayer 2006; Marks and Smith 2006). The NGDC tsunami-runup database can be searched using the cause of the tsunami parameter set to the mechanism of interest. Tsunami runups that are attributed to landslides sometimes contain data regarding the landslide properties, including extent and volume.

The NGDC also archives a grain-size database of sea-floor sediments. The database contains over 17,000 sea-floor samples from the Outer Continental Shelf Environmental Assessment Program (OCSEAP), academic institutions, and the U.S. Navy. The metadata include the name of the collecting institution, the ship, the cruise, the sample identifier, the location, the date of collection, the water depth, the sampling device, the method of analysis, the weight of the sample, the sampling interval, and the raw-weight percentages. Some of the samples also have texture information (percentage of sand, silt, and clay) and statistical measures of grain size, such as mean, median, standard deviation, skewness, and kurtosis of the grain-size distribution. The NGDC grain-size database can be interactively searched at the website http://www.nadc.noaa.gov/mag_grainsize/index.isp. For example, a search off the West Coast of the United States, bounded by a rectangle with latitudes ranging from 32°N to 44°N and longitudes ranging from 122°W to 130°W, returned four matching cruises and 22 samples.

4.8. Paleotsunami Data

Tsunamis deposit sediment in the runup zone. These deposits can be overlain by sediments brought to the site by other mechanisms and, over a long period of time, may get buried in the soil profile. The NGDC is compiling a database of paleotsunami data (see Appendix A of Gonzalez et al. 2007). However, no central repository of paleotsunami data exists at this time.

Prehistoric tsunamis have been identified based on the deposits in the Pacific Northwest (Atwater and Moore 1992; Benson et al. 1997; Peters et al. 2003), in Kamchatka (Pinegina and Bourgeois, 2001; Pinegina et al. 2003), in Japan (Nanayama et al. 2003), in the North Sea (Dawson et al. 1988), and in Hawaii (Moore 2000; Moore et al. 1994a, 1994b). Peters et al. (2003) compiled a database of deposits related to tsunamis in the Cascadia. This database contains the location and sedimentological properties from 59 sites located in the Pacific Northwest, starting in northern California and extending to Vancouver Island, British Columbia. The database references 52 published studies until the year 2002 that describe these tsunami deposit sites. The individual studies cited in the report use several different criteria to distinguish tsunami deposits, which usually manifest as anomalous sand layers in coastal mashes and lacustrine sediments, from sand layers deposited by other processes, such as river flooding or storm surges.

Preliminary identification of tsunami deposits is carried out by searching for layers of sand or coarse-grained material in outcrops and sediment cores from environments where deposition of sand layers is unusual, such as in coastal marshes and in lakes (Peters et al. 2003). Tsunamis can transport a variety of sizes of sediment, ranging from sand particles to large cobbles and boulders. Sand layers can also be deposited by other processes, such as storm surges or river flooding, in these environments. Usually, a combination of key factors related to the deposited layer are used to distinguish it as a tsunami deposit. These key characteristics include biological markers, spatial distribution of the deposits, sediment characteristics, and geochemistry. The presence of marine macro and microfossils is used to infer a marine source for the sediment layer. Progressively inland thinning and finning of the deposited layer is often used to suggest a marine surge rather than river flooding as the mechanism responsible for these deposits. Composition and texture of sand grains can be compared with upriver deposits to distinguish the mechanisms responsible for these deposits. Geochemical indicators, such as bromine enrichment, have also been used to indicate marine source of the deposits.

Usually, it is more difficult to distinguish storm- surge deposits from those due to tsunamis because both contain marine macro- or microfossils, have saltwater chemistry, and progressively thin and fine inland from the shore. The distance from the shore of the deposits, the presence of several relatively thick normally graded[(a)] layers that may indicate several high-energy tsunami waves, and the presence of rip-up clasts that also indicate high energy of the tsunami waves are used to distinguish deposits of tsunami origin.

Association of the sand layer with paleoseismicity is also used to link a sand layer to a tsunami source. Coseismic coastal subsidence may accompany great earthquakes on subduction zones (Atwater 1996). A buried tsunami sand layer may be deposited over a layer of marsh peat, which is then overlain by tidal mud after subsidence.

The database compiled by Peters et al. (2003) contains data on the age, number of deposits, sedimentary characteristics, and identifying features of the Cascadia tsunami deposits.

Although the focus of the database is on tsunamis originating in the Cascadia subduction zone, data related to transoceanic tsunamis have also been included. However, the authors stated that tsunami deposits located in the Puget Sound area that are not believed to be of Cascadia origin were excluded.

The database is presented as a spreadsheet and a geographic information system coverage. The database includes the following information related to each reported deposit:

1. location: the name of the place where the tsunami deposit is located, including the name of the state
2. catalog number: an arbitrary serial number
3. site number: a number assigned to the deposits, in ascending order with latitude
4. core/section/secondary location: individual cores from a site that are listed separately are identified using a core number, or the section or secondary location number used by the referenced study
5. latitude and longitude
6. depositional setting: such as a lake, coastal marsh, freshwater marsh, etc.

(a) Normal grading of a sediment layer refers to upwards fining in contrast to inverse grading that refers to an upwards coarsening of sediments.

7. physiographic setting: such as head of inlet, coastal lake or marsh, etc.
8. inundation distance: inland from the shore
9. inundation reference: such as open coast, upriver, etc.
10. elevation above mean sea level
11. barrier elevation: elevation of a barrier that the tsunami must have crossed to lay the deposit
12. observation/sampling method: method used to collect the data, such as observation from and outcrop, the type of coring device used, etc.
13. number of cores/sampling localities at a site
14. number of subsidence events associated with great earthquakes on the Cascadia subduction zone, irrespective of their association with tsunamis
15. number of tsunami events documented at the site
16. whether coseismic subsidence was present
17. number of tsunami deposits at a site that are associated with coseismic subsidence
18. number of tsunami deposits at a site that are not associated with coseismic subsidence
19. event number referring to the tsunami or subsidence, from youngest to oldest
20. tsunami event number, from youngest to oldest
21. subsidence event number, from youngest to oldest
22. amount of subsidence
23. age of deposition in radiocarbon years before present (usually reported relative to 1950 A.D.)
24. age range: the range of possible ages in calendar years before present (present taken as 1950 A.D.)
25. correlated date: the date of the event the deposit is correlated to by the author, or otherwise accepted
26. method used to determine the age of deposits
27. thickness of the tsunami deposit
28. maximum thickness
29. geometry: gradient, landward thinning, and continuity of the deposit
30. number of tsunami deposit layers
31. layer thickness
32. layer characteristics, such as texture, grading, etc.
33. underlying and overlying material
34. lower and upper contacts that bound the deposit
35. grain size range, distribution, and description

36. textural gradient in the horizontal direction, such as landward fining or fining away from the channel
37. textural gradient in the vertical direction
38. sorting, which is a measure of the variability of grain sizes in the deposit
39. other sedimentary properties described by authors of the original report
40. minerological composition of sediment grains, such as presence of quartz, feldspar, or lithics
41. inclusions, such as plant material, shells, microfossils, artifacts, etc.
42. flow direction as indicated by the deposit
43. microfossils, such as forams and diatoms
44. chemical evidence
45. additional description pertaining to the deposits, the site, or publications that do not fit other categories
46. reference, including type and date of publication
47. a reference map of the location.

At this time, no similar sources of paleo-tsunami data are known that describe other regions of the United States. It may be necessary to collect site-specific data using the same methods and approaches described above. The list of characteristics given above may be used as a starting point of the inventory of site-specific paleotsunami characterization.

5. Probable Maximum Tsunami

5.1. Introduction

Consideration of tsunamis in determination of safety of a nuclear power plant is required by Title 10 of the Code of Federal Regulations, Part 50, Appendix A, General Design Criterion 2, Design bases for protection against natural phenomena, which states:

> Structures, systems, and components important to safety shall be designed to withstand the effects of natural phenomena such as earthquakes, tornadoes, hurricanes, floods, tsunami, and seiches without loss of capability to perform their safety functions. The design bases for these structures, systems, and components shall reflect: (1)

> Appropriate consideration of the most severe of the natural phenomena that have been historically reported for the site and surrounding area, with sufficient margin for the limited accuracy, quantity, and period of time in which the historical data have been accumulated, (2) appropriate combinations of the effects of normal and accident conditions with the effects of the natural phenomena and (3) the importance of the safety functions to be performed. (Emphasis added.)

The PMT is not estimated using a probabilistic approach. It is, on the contrary, a deterministic approach that incorporates ideas of transposition and maximization, similar to the methods adopted by the National Oceanic and Atmospheric Administration National Weather Service Hydrometeorological Reports for estimation of Probable Maximum Precipitation (Schreiner and Riedel 1978). At this time, sufficient observed data regarding tsunamis, particularly large tsunamis, and their effects are not available to support a comprehensive probabilistic tsunami-hazard analysis or PTHA (e.g., see Geist and Parsons 2006) at all probable nuclear power plant sites in the United States to arrive at a set of design bases. In the future, as more data are collected and scientific methods advance, PTHA may become a viable tool to assess tsunami hazards and to specify these design bases.

A complete PMT assessment is only required if the hierarchical-hazard-assessment approach described in Chapter 2 of this report results in a determination that the site region and the site are subject to tsunami hazards, or that the result of the hierarchical assessment is inconclusive that the site is not subject to hazards from a tsunami.

5.2. Definition

For the purposes of this report, we have adopted the following definition of the PMT:

> PMT is that tsunami for which the impact at the site is derived from the use of best available scientific information to arrive at a set of scenarios reasonably expected to affect the nuclear power plant site, taking into account (1) appropriate consideration of the most severe of the natural phenomena that have been historically reported for the site and surrounding area, with sufficient margin for the limited accuracy, quantity, and period of time in which the historical data have been

accumulated; (2) appropriate combinations of the effects of normal and accident conditions with the effects of the natural phenomena; and (3) the importance of the safety functions to be performed.

The salient points of this definition of a PMT are (1) it allows use of the best available scientific methods and data, (2) it accounts for the limited period of time in which tsunami data have been collected in the United States, (3) it allows for sufficient margin in design bases to account for the limited accuracy and quantity of observed data, and (4) it allows for accounting of the importance of the safety functions to be performed by the affected SSC.

5.3. Determination of Probable Maximum Tsunami at a Nuclear Power Plant Site

The Standard Review Plan (USNRC 2007) describes PMT hazards and the approach for determination of safety of a nuclear power plant. The areas of review for a nuclear power plant siting include historical tsunami data, the PMT including relevant source mechanisms, tsunami propagation and inundation models, wave runup, inundation, and drawdown, hydrostatic and hydrodynamic forces, debris and projectiles, and effects of sediment erosion and deposition. The following subsections describe the approaches and methods for determination of the PMT and hazards posed by it at the nuclear power plant site.

5.3.1. Tsunamigenic Mechanisms and Sources

For determination of the PMT, three mechanisms should be considered: earthquakes, landslides, and volcanoes. Regardless of a site being coastal or inland, if it is located near a water body that can support generation of a tsunami or a tsunami-like wave triggered by any of the three mechanisms, a PMT assessment should be carried out.

Because both near- and far-field sources should be considered for determination of the PMT, the search area for tsunamigenic mechanisms can be quite large, especially for sites located on the coasts. This search should review all available sources of historical and prehistorical data and create a list of tsunamigenic sources for all mechanisms that may be relevant for the nuclear power plant site.

It may not be possible to a priori determine which tsunamigenic source may generate the PMT. Several candidate sources and the tsunamis generated from them may require evaluation under the most favorable tsunamigenic source and ambient conditions. Deterministic tsunami propagation and inundation modeling can be carried out with appropriately conservative sets of source parameters to evaluate the final PMT for a given nuclear power plant site.

5.3.2. Source Parameters

For determination of the PMT, conservative values and ranges of source parameters should be specified. This ensures that the design bases of the nuclear power plant will not be exceeded. A comprehensive search of historical and prehistorical records should be undertaken to identify potential tsunamigenic sources relevant to the nuclear power plant site. It is possible that for some tsunamigenic mechanisms and sources, a corresponding historical or prehistorical tsunami may not be identified, in part due to the relatively short period of record. However, for such mechanisms and sources, their maximum tsunamigenic parameters may be estimated and a deterministic hydrodynamic modeling used to assess their potential impact at the nuclear power plant site.

Determination of source parameter values associated with the three mechanisms may require expert opinion supplemented with historical case studies and available historical data. An example of such a study is given by Gonzalez et al. (2003).

The following source parameters for each of the source mechanisms should be determined.

Earthquakes

The source parameters for a tsunamigenic earthquakes are

1. location (latitude, longitude, and depth)
2. moment magnitude,
3. fault dimensions (rupture length and width)
4. dip
5. strike
6. slip distribution
7. shear modulus.

Landslides

The source parameters for a submarine landslide are

1. slide location (latitude and longitude)
2. slide orientation (directivity)
3. slide volume
4. slide speed and acceleration
5. cohesiveness of the slide material.

The source parameters for a subaerial landslide, including ice falls, in addition to those listed above for submarine landslide are

1. density of slide material
2. impact Froude number.

Volcanoes

The source parameters for a tsunamigenic pyroclastic flow are

1. location
2. orientation
3. density of the flow material
4. dimensions of the flow
5. impact Froude number.

The source parameters for a caldera collapse are

1. location
2. orientation
3. caldera dimensions
4. fixed wall height
5. cavity depth after the collapse.

The source parameters for a submarine volcanic explosion are

1. location
2. volcanic Explosivity Index.

The source parameters for tsunamigenic-debris avalanches and flank failures are similar to those for submarine and subaerial landslides.

5.3.3. Initial Waveform

The tsunamigenic mechanism results in the displacement of the water surface, which is the initial tsunami waveform. This initial waveform evolves hydrodynamically into the train of waves that propagates from the source toward the site.For short-duration mechanisms (i.e., when the duration of source dynamics is much smaller than the period of the tsunami waves), the coupling between the source and the initial waveform can be neglected without significantly affecting the properties of the tsunami waves. This approach is suitable for earthquake-generated tsunamis. The formulation of:Mansinha and Smylie (1971), later modified by Okada (1985), based on an elastic earth crust may be used to specify the initial waveform. The Okada formulation is a closed-form analytical expression and applies to point and finite rectangular sources, suitable for computing an average slip for the entire fault plane. For more complex ruptures, a slip-distribution model may be needed. One way to construct a complex rupture pattern is to divide the fault plane in a series of smaller faults, each of these individually descry bed by an Okada formulation. Usually, the displacement pattern of the bottom is assumed to also apply to the water surface and, therefore, also specifies the initial waveform of the tsunami.

The velocity of landslides can often be comparable to the phase velocity of the tsunami waves generated by it. Explicit landslide models should be employed in such cases (e.g., Titov and Gonzdlez 2001). Some examples of landslide models applied for tsunami generation are described by Jiang and LeBlond (1992), Watts (1998), and Fine et al. (1999).

Because the volcanic mechanisms that generate tsunamis are often similar to submarine and subaerial landslides, explicit models of initial waveform generation from volcanic source mechanisms may be needed.

5.3.4. Wave Propagation Simulation

Oceanic tsunamis

Tsunamis generated by large earthquakes have long wavelengths due to the large spatial scale of the seismic source. The tsunami wavelengths can range from tens to hundreds of miles. The tsunami wavelength is large compared to the depth of the water [the deepest point in Earth's oceans is the Mariana Trench in the Pacific at 6.8 mi (10.9 km)], and, therefore, tsunami waves are called long waves or shallow water waves, the latter term referring to the shallowness of the water compared to the wavelength.

Although closed-form expressions are available that describe some properties of the tsunami (e.g., the propagation speed of the tsunami in deep ocean, which can be used to approximately determine the time of arrival of a far-field tsunami at a given site), detailed behavior of tsunami-wave dynamics can only be determined using numerical simulations.

Two community numerical models are available: (1) the Method of Splitting Tsunami (MOST) developed originally at University of Southern California (Titov 1997; Titov and Synolakis 1997) and since implemented, maintained, and improved at NOAA's Pacific Marine Environmental Laboratory (Titov and Gonzalez 1997); and (2) the set of models (TSUNAMI-N1, TSUNAMI-N2, and TSUNAMI-N3 for simulation of near-field tsunamis and TSUNAMI-Fl and TSUNAMI-F2 for simulation of far-field tsunamis) developed by Imamura and colleagues (Imamura et al. 2006). These models have the advantage of widespread use in the tsunami-modeling community and have benefitted from comparisons with observed tsunami data and subsequent updates to the model formulations.

Summary of the MOST model

The MOST model is a suite of numerical programs that simulate all three phases of the tsunami: generation by an earthquake, propagation across the ocean, and runup.

Tsunami generation is simulated using the fault plane model of Okada (1985). This model assumes an incompressible water layer overlies an elastic half space that represents the crust of the earth.

The MOST model uses non-linear shallow-water wave equations, including Coriolis terms, expressed in a spherical coordinate system. Dispersion of the tsunami waves, which is an effects of the dependence of wave celerity on the frequency of component waves, is handled by taking advantage of the numerical dispersion inherent in a finite difference scheme as suggested by Shuto (1991). This dispersion scheme allows MOST to use non-dispersive governing equations. The governing equation are (Titov and Gonzalez 1997)

$$h_t + \frac{(uh)_\lambda + (vh\cos\phi)_\phi}{R\cos\phi} = 0$$

$$u_t + \frac{uu_\lambda}{R\cos\phi} + \frac{vu_\phi}{R} + \frac{gh_\lambda}{R\cos\phi} = \frac{gd_\lambda}{R\cos\phi} + fv$$

$$v_t + \frac{uv_\lambda}{R\cos\phi} + \frac{vv_\phi}{R} + \frac{gh_\phi}{R} = \frac{gd_\phi}{R} - fu$$

where λ denotes the direction along the longitude, φ denotes the direction along the latitude, t denotes time, h(λ, φ, t) is the amplitude of the wave, d(λ, φ, t) is the depth of undisturbed water, u(λ, φ, t) is the depth-averaged velocity in the direction of the longitude, v(λ, φ, t) is the depth-averaged velocity in the direction of the latitude, R is the radius of the earth, g is the acceleration due to gravity, $f=2\omega\sin\varphi$ is the Coriolis parameter, ω is the angular velocity of the earth, and subscripts λ, φ, and t denote partial derivatives with respect to those space and time dimensions. A more detailed description of the model can be found in Gonzalez et al. (2007).

Summary of the Immamura models

Two general classes of tsunami models, one for near-field tsunamis (TSUNAMI-N1, TSUNAMI-N2, and TSUNAMI-N3) and the other for far-field tsunamis (TSUNAMI-PI and TSUNAMI-F2) were described by UNESCO (1997) and Imamura et al. (2006).

The near-field tsunami models use a cartesian coordinate system, and the far-field tsunami models use a spherical coordinate system. Both sets of models are based on the non-dispersive, nonlinear, shallow-water wave-governing equations similar to that used by the MOST model; the far-field models also incorporate the Coriolis terms in their governing equations. Both sets of models use a finite-difference discretization scheme to solve the governing equations over a computational grid.

TSUNAMI-N1 uses linear approximation of the nonlinear shallow-water wave equations with a constant computational grid. TSUNAMI-N2 uses linear approximation in deep waters and the shallow-water wave equations in shallow waters and for computing runup, both over a constant computational grid. TSUNAMI-N3 uses linear approximation of the nonlinear shallow-water wave equations with a varying computational grid.

TSUNAMI-F1 model uses linear approximation of the nonlinear, shallow-water wave equations to simulate propagation of the tsunami in deep ocean using a spherical coordinate system. TSUNAMI-F2 model uses linear equations, but also includes coastal runup simulations.

Tsunamis in other water bodies

Except for the very last stages of runup on the shore, when the tsunami wave steepens to such an extent that breaking may occur, the tsunami can be considered a long wave (Titov and Synolakis 1997). Therefore, tsunami-like waves in other bodies of water may also be simulated using the models described above. The main difference between oceanic tsunamis and tsunami-like waves in other water bodies may be the source mechanism. Landslides and hillslope failures may be more likely to generate tsunami-like waves in water bodies other than oceans. Under these circumstances, a source model may be needed to describe the initial tsunami generation from landslides, hillslope failures, and debris avalanches (i.e., subaerial landslides). The initial waveform predicted by these source models may be used to specify the initial conditions for the tsunami propagation and inundation models, such as those described above.

5.3.5. The NOAA Center for Tsunami Research Tsunami Propagation Database

Characteristics of far-field tsunamis generated by an earthquake source depend mainly on a few source parameters, like the location and magnitude—assuming some typical displacement mechanism (Titov et al. 1999; Gica et al. 2006). Also, linear approximation of the shallow-water wave equations has been found to be applicable in deep ocean. Therefore, a set of unit sources—located along the tsunamigenic subduction zones can be used to construct a tsunami-propagation scenario from an earthquake of given magnitude and a given location by linearly combining the numerical solution from all unit sources—is needed to describe an earthquake of interest.

This approach was used to construct a forecast database of precomputed propagation solutions for unit sources around the Pacific and a preliminary database in the Atlantic. Details of the definition of these unit sources is described by Titov et al. (1999; 2005) and Gica et al. (2006). The source parameters for the unit sources in the Pacific are specified using the latest estimates of these parameters in the corresponding subduction zones (Kirby et al. 2006). A similar effort is also underway to define unit sources in the Atlantic that are of relevance to the U.S. coastline.

One potential use of the NOAA Center for Tsunami Research (NCTR) tsunami-forecast database is to provide boundary conditions to near-shore inundation models. This approach of separating the inundation modeling (which is site dependent) and the modeling of generation and propagation is an attractive approach. The NCTR tsunami database can provide robust and verified estimates of tsunami propagation. Near-shore data collection for parameterization of a tsunami inundation model and estimation of hazard matrices can be carried out on case-by-case, site- specific requirements. However, this approach is available only for earthquake sources in the Pacific and is being developed for Atlantic sources. Submarine landslide and volcanic sources are not treated similarly. Also, this approach may not work for inland water bodies.

5.4. Hazard Assessment

The effects of a tsunami at a nuclear power plant site are described in Chapter 3 of this report. In this section, hazards, and specifically metrics that describe these hazards and can be used in the specification of design bases of the nuclear power plant, are described.

The effects of tsunamis on the shoreline can be highly variable depending on the characteristics of the tsunami, including its directivity and wavelength, the interaction of the tsunami waves with offshore and near-shore bathymetry, and the geotechnical characteristics of the near-shore substrate. Due to these geospatial variations and the highly nonlinear behavior of the tsunami waves near the shoreline, the severity of various hazards from different tsunamis may be different. Consequently, the most severe occurrence of a particular tsunami hazard at a site may result from a set of tsunamis, not from a particular tsunami. Because the goal of the tsunami-hazards assessment is to determine the most severe hazard at a site, hazard assessment from a set of PMTs may be required. Therefore, tsunami-hazard assessment should consider a set of candidate PMTs for which hazards at the site should be determined. The hazard that is the most severe among this set of candidate PMTs should be used.

5.4.1. High Water Level

Flooding is the most obvious of the hazards from a tsunami. The runup, which is defined as the maximum ground elevation that the tsunami waves reach above a standard datum, is the metric used to define the high-water level. Ambient water conditions, such as high tide, can affect the high- water level.

The runup is obtained as an output from the inundation model. Significant progress has been made in the recent past to improve the inundation models, especially with regard to validation of model predictions with observed runup data (Titov and Synolakis 1997). Model predictions of runups can be in disagreement with observations, often by a large margin and especially with regard to the maximum observed runup along the coastline (Titov and Synolakis 1997). Some of the errors may be attributed to insufficient resolution of local bathymetric and topographic data.

Along with the maximum runup, the areal extent of the tsunami runup should also be considered. The uncertainties in model prediction of runup also extend to the prediction of areal extent of flooding during tsunami inundation.

Due to significant uncertainty that still exists in predictions from inundation models, care should be taken in application of these models and use of their predictions to specify design bases with respect to the locations of SSC important to safety as well as their grade elevations. Where possible, inundation models should be verified with locally available runup data. Sufficient margins should be provided for these uncertainties in all design bases that may be derived from such simulations.

5.4.2. Low Water Level

Low-water level can lead to a dry intake, compromising the safety of a nuclear power plant if the intake is used to provide safety-related cooling water. The drawdown, which is defined as the minimum water-surface elevation caused by the tsunami waves at the shoreline, is the metric used to describe low-water level. Ambient conditions, such as low tide, can affect the low-water level.

The drawdown, similarly to the runup, is predicted by the inundation model. Because of uncertainty inherent in the predictions from the inundation models, arguments similar to those for runup also apply to the drawdown. In addition, it is more difficult to obtain accurate measurements of drawdown caused by a tsunami because these locations are under water during normal conditions. Usual techniques, like high-water marks or elevation of debris lines used to infer runups during post-tsunami surveys, are not applicable. Accurate measurements of drawdown mainly relies on eyewitness accounts, photographs, or satellite imagery. Therefore, validation of drawdown predictions from inundation models is more difficult than that for runups.

Due to the reasons described above, care should be taken in interpretation of drawdown predictions from inundation models, and sufficient margins should be provided in all design bases that may be affected by these predictions.

5.4.3. Scouring Near Safety-Related Structures

Tonkin et al. (2003) proposed a scour enhancement parameter Λ, defined as the fraction of the buoyant weight of the sediment grains that is supported by the pore-water pressure gradient. They provided an expression for Λ as a function of the depth of the substrate given

$$\text{as } \Lambda(d_s) = \frac{\Delta P}{\gamma_b d_s}\left(1 - 4i^2\text{erfc}\left[\frac{d_s}{2\sqrt{c_v \Delta T}}\right]\right)$$

where γ_b is the buoyant specific weight of the saturated soil, d_s is the movable soil depth, c_v is the coefficient of consolidation, ΔP is the peak surface pressure that decreases linearly to zero over the time period ΔT, and $i^2\text{erfc}[\bullet]$ is the second integral of the complementary error function. Tonkin et al. (2003) found that the scouring occurred to a depth such that the estimated value of Λ was 0.5.

The condition that any enhanced scour may take place is obtained by letting $d_s \rightarrow 0$, which yields the following expression (Tonkin et al. 2003):

$$\Lambda(0) = \frac{2}{\sqrt{\pi}} \frac{\Delta P}{\gamma_b \sqrt{c_v \Delta T}}$$

The expressions above may be used to estimate if any enhanced scouring near safety-related structures is expected to occur during tsunami inundation and drawdown. The foundations of safety-related structures should be designed to protect against scouring. An alternative option may be to site all safety-related structures in areas that are not exposed to enhanced scouring caused by tsunami waves.

5.4.4. Deposition Near Safety-Related Structures

No models of sediment transport during tsunamis have been established. On a case-by-case basis, site-specific determination of sediment transport and accumulation, and their effects on SSC important to safety, should be undertaken. All SSC important to safety should be sited such that sediment accumulation would not result in loss of their functionality.

5.4.5. Forces on Safety-Related Structures

Hydrostatic force

The hydrostatic force, f_h, per unit width on a wall that is not overtopped by the tsunami waves is given by Yeh et al. (2005):

$$f_h = \frac{1}{2}\rho g\left(h + \frac{u_p^2}{2g}\right)^2$$

where ρ is the density of water, g is the acceleration due to gravity, h is the water depth, and u_p is the component of water velocity normal to the wall. The force acts horizontally at a distance h_R above the base of the wall, with h_R given by the follow expression (Yeh et al. 2005):

$$h_R = \frac{1}{3}\left(h + \frac{u_p^2}{2g}\right)$$

Notice that the hydrostatic force also includes a velocity head component, in contrast with traditional definition of hydrostatic forces. The velocity head is included because inundating water during tsunami runup will typically have a significant velocity component.

Hydrodynamic force

The parameters of tsunami flow that are important to assess dynamic effects (González et al. 2007) are total flow depth,

$$h = d + \eta,$$

where d is the still water depth and η is the local amplitude of the tsunami; the tsunami current speed,

$$V = \sqrt{u^2 + v^2},$$

where u and v are velocity components in the two horizontal directions in a reference coordinate system; the acceleration of tsunami current in the direction of the flow,

$$\frac{dV}{dt},$$

where t refers to the time dimension; the inertial component,

$$h\frac{dV}{dt},$$

and the momentum flux,

$$hV^2$$

Kanoglu and Synolakis (2006) investigated the evolution of two simple waveforms on a simple sloping beach and found that even for wave runup over simple geometries, the points of maximum depth of inundation and maximum flow velocity are not the same. They also found that the location of the point of maximum velocity depends on the incoming waveform. Therefore, care should be exercised in interpreting a few dynamic parameters estimated from inundation models.

The hydrodynamic force, or the drag force, F_d, on a structure in the tsunami flow field is given by Yeh et al. (2005):

$$F_d = \frac{1}{2}\rho C_d A u_p^2$$

where C_d is the coefficient of drag and A is the projected area of the structure on a plane normal to the flow direction. Appropriate values of the coefficient of drag are provided by FEMA (2005) and reproduced here in Table 5-1.

Table 5-1. Drag coefficients for ratios of width to depth (w/ds) and width to height (w/h) [adapted from FEMA (2005)]

Ratio (w/d or w/h)	C_d
1-12	1.25
13-20	1.3
21-32	1.4
33-40	1.5
41-80	1.75
81-120	1.8
>120	2.0

Wave forces

Breaking wave forces on vertical piles and columns is given by FEMA (2005):

$$F_{bkrp} = \frac{1}{2} \rho g C_{db} D H_b^2$$

where **D** is the diameter of the pile or column, H_b is the breaking-wave height, and C_{db} is the breaking wave drag coefficient. FEMA (2005) recommends $H_b = 0.78\ d_s$ where d_s is the design stillwater-flood depth. The recommended value of C_{db} is 2.25 for square or rectangular piles and columns and 1.75 for round piles and columns (FEMA 2005).

Breaking-wave forces on vertical walls is given by FEMA (2005) based on two cases: (1) where the wave breaks against a wall that encloses dry space behind it, and (2) where the stillwater level is same on both sides of the wall. The estimation equation for f_{brkw}, the total breaking wave force per unit width of the wall, is

$$f_{brkw} = \begin{cases} (1.1\ C_p + 2.41) \rho g d_s^2 & \text{for case 1} \\ (1.1\ C_p + 1.91) \rho g d_s^2 & \text{for case 2} \end{cases}$$

where C_p is the dynamic pressure coefficient (Table 5-2 below).

5.4.6. Debris Accumulation

No models have been established for debris accumulation during tsunamis. On a case-by-case basis, site-specific determination of debris transport and accumulation, and their effects on SSC important to safety, should be undertaken. All SSC important to safety should be sited such that debris accumulation would not result in loss of their functionality.

Table 5-2. Value of Dynamic Pressure Coefficient, C_p, as a Function of Probability of Exceedance [Adapted from FEMA (2005)]

C_p	Building Type	Probability of Exceedance
1.6	Accessory structure, low hazard	0.5
2.8	Coastal residence	0.01
3.2	High-occupancy building or critical facility	0.001

Table 5-3. Impact Durations for Selected Construction Materials [Adapted from FEMA (2005)]

Type of Construction Material	Impact Duration (sec)	
	Wall	Pile
Wood	0.7-1.1	0.5-1.0
Steel	Not applicable	0.2-0.4
Reinforced concrete	0.2-0.4	0.3-0.6
Concrete masonry	0.3-0.6	0.3-0.6

5.4.7. Projectiles

The impact force, F_I, on the structures at the water level is given by Yeh et al. (2005):

$$F_I = m\, a_b = m\frac{du_b}{dt} = m\frac{u_I}{t_I}$$

where m is the mass of the impacting projectile, a_b is the acceleration of the impacting projectile, u_b is the velocity of the impacting projectile, u_I is the approach velocity assumed equal to flow velocity, and t_I is the duration of impact. FEMA (2005) provides durations of impact for different construction materials (Table 5-3).

Large objects transported by the tsunami waves, such as boats, may turn into projectiles impacting a structure at a height above the surface of the water level. These potential impacts should be studied carefully to specify the impact location on the structure and appropriately modify the design bases. The assumption that debris velocity is equal to flow velocity may be accurate for small objects, but may overestimate the debris velocity of large objects (FEMA 2005).

5.5. Combined Effects

At the time a tsunami arrives at the nuclear power plant site, ambient conditions may affect its properties. For example, at a coastal site, tides can affect the ambient highest and lowest water levels. From a design perspective, a low-tide level (e.g., the 90% exceedance low tide) should be combined with the drawdown from the tsunami to arrive at the low-water level. Similarly, a high-tide level (e.g., the 10% exceedance high tide) should be combined with the runup to arrive at the high- water level.

In the presence of offshore submerged currents, the effects of these currents on the propagation and inundation of tsunami waves should be considered. It may be desirable to include the effects of such currents in the tsunami propagation and inundation models themselves to account for any dynamic interaction between the current-flow field and the tsunami-wave field.

For near-field earthquake-generated tsunamis, the effects of the earthquake may need to be considered in combination with those of the tsunami. For example, ground motion due to the earthquake may be an additional precursor to a potential liquefaction of the soil near the foundation of safety-related structures that are subsequently inundated and scoured by tsunami waves. Under such circumstances, the stresses from the earthquake ground motion should be considered in addition to the buoyant forces due to the tsunami- wave action to arrive at the design-basis liquefaction potential.

The effects of a tsunamigenic event to trigger another tsunamigenic event should also be considered. For example, a severe earthquake may generate a tsunami by vertical displacement of the water column in a water body and may also trigger a subaqueous landslide that may generate a separate tsunami. The two tsunamis may combine constructively to produce far larger tsunami events. Some researchers suggest that these concomitant tsunamigenic events may explain some of the so-called tsunami earthquakes where the generated tsunami is abnormally large compared with the size of the tsunami expected based on the earthquake magnitude alone.

6. International Practices

6.1. Introduction

This section presents a review of accepted international practices for tsunami hazard assessment at nuclear power plant sites.

6.2. Japan

Tsunami-hazard assessment is a necessity for nuclear power plant sites in Japan. Consequently, Japanese tsunami-hazard-assessment approaches are some of the most advanced in the world. The assessment method for the tsunami hazard at Japanese nuclear power plants is described by the Japanese Society of Civil Engineers (JSCE) (2002).

The Japanese assessment approach only describes earthquake-induced tsunamis. Volcano- and meteorite-induced tsunamis are considered infrequent events in comparison with earthquakes and are excluded from consideration. The hazard caused at the nuclear power plant site is also limited to estimation of high- and low-water levels caused by the design tsunami.

The design tsunami is defined as one that causes the maximum water rise or fall at the nuclear power plant site. The design water level is defined as the sum of water level caused by the design tsunami in combination with an appropriate tidal condition.

Even though Japan has the most extensive historical database of tsunamigenic earthquakes, significant uncertainty in source parameters exist. To account for this uncertainty, JSCE requires a parametric study, varying the source parameters within a reasonable range to numerically simulate a number of scenario tsunamis. The tsunami or tsunamis that cause the maximum water rise or fall at the site are selected as design tsunamis.

Design tsunamis are verified by comparing the water level it produces with the water levels corresponding to all recorded and numerically simulated historical tsunamis at the site. Additionally, the envelope of scenario tsunami-water levels in the vicinity of the site should exceed all recorded or simulated historical tsunami-water levels. An illustration of the design tsunami-concept is shown in Figure 6-1. The numerical models used for simulation of scenario tsunamis to estimate water levels at the site are verified using historical tsunami observation.

Scenario tsunamis (thin black curves in Figure 6-1) may produce maximum or minimum water levels at points different from where the site is located along the coastline in response to directivity of the earthquake source and bathymetry along the wave-propagation path from the source to the coastline that modifies the waveform. The scenario tsunami that results in the maximum of minimum water level at the site (thick black line in Figure 6-1) is the design tsunami. Verification of the design tsunami is based on two criteria: (1) the design tsunami should produce a more severe water level than all historically recorded tsunami water levels and all numerically inferred water levels from historical known tsunamis at the site (thick black curve enveloping vertical bars at the site in Figure 6-1), and (2) the envelope of scenario tsunamis (thick red line enveloping all high water levels produced by scenario tsunamis in Figure 6-1) should produce more-severe water levels than all historically recorded tsunami water levels and all numerically inferred water levels from historically known tsunamis in the vicinity of the site. Both near- and far-field sources are considered for determining design tsunamis.

Standard fault model and scenario earthquakes

A standard fault model is defined as an appropriate earthquake model for a site that is used in numerical tsunami simulations. The standard fault model is chosen based on careful examination of historical tsunamigenic earthquakes and other information regarding seismic activity that are relevant for a given site.

The fault motion is modeled using a rectangular, uniform slip fault plane that is described using nine parameters:

1. latitude of the reference point
2. longitude of the reference point
3. Depth of upper edge of the fault plane
4. fault length
5. fault width
6. slip amount
7. strike direction
8. dip angle
9. slip angle.

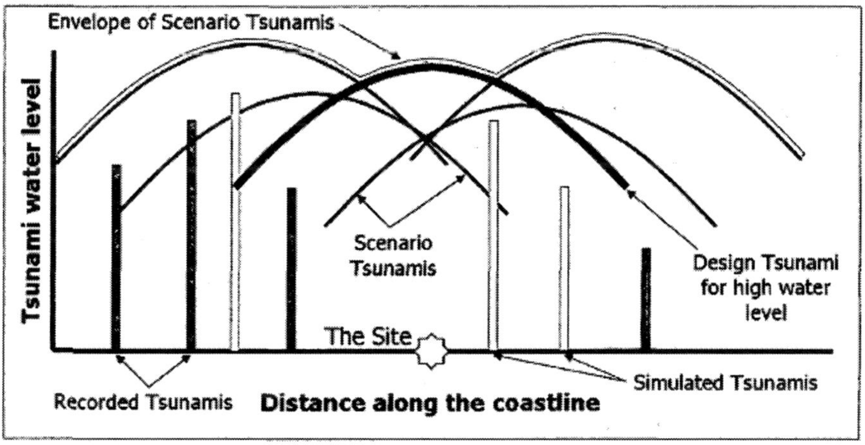

Figure 6-1. Conceptual illustration of design tsunami (JSCE, 2002), adapted with permission.

Scenario earthquakes are specified by varying their respective fault motion parameters within a reasonable range of the baseline parameter set, that of the standard fault model. Fault motion parameters whose variations can be probabilistically quantified are varied one standard deviation from their baseline, standard fault model value.

Numerical tsunami simulation

The governing equations used in numerical tsunami simulations are depth-averaged hydrodynamic equations that describe the tsunami wave propagation in two dimensions. The linear version of the governing equations may be used when the ratio of wave height to water depth is sufficiently small. Nonlinear equations should be used when the ratio of wave height to water depth is not small and nonlinearity of the governing equations cannot be ignored. The equation of motion in this case includes an unsteady term, a pressure term, a bottom-friction term, and an advection term that facilitate modeling of the steepening of the wave front as it propagates in shallower waters. A dispersion term may also be needed if the curvature of the tsunami waves increases during propagation.

Generally, for simulation of nearshore-tsunami propagation, when the water depth is less than 656 ft (200 m), nonlinear governing equations are employed. The recommendation is to use an explicit finite difference scheme with a staggered leapfrog method for discretization of the governing equations. Two methods are specifically recommended: the Goto and Tanaka methods. Both methods use the staggered leapfrog scheme. The Goto method uses a conservation-type advection term, whereas the Tanaka method uses a nonconservation-type advection scheme. Friction term in Goto method is based on a Manning approach, whereas the Tanaka method uses a general friction term. Horizontal eddy viscosity term is used in the Tanaka method, but is introduced in the Goto method if thought necessary. The Goto method uses a first-order upstream-difference scheme with associated accuracy of first order for the advection term, whereas the Tanaka scheme uses the Lax-Wendroff scheme with second-order accuracy.

The depth-averaged two-dimensional hydrodynamic equations may not be appropriate for use in runup simulation if the local topography near the maximum runup location possesses steep slopes of a small valley. These conditions were encountered near the village Monai on Okushiri Island during the Hokkaido Nansei-Oki tsunami. In such cases, a model based on nonlinear three- dimensional governing equations may become necessary.

Far-field propagation of a tsunami can be conducted using the linear version of the governing equations because the wave height is small compared to water depth during transoceanic propagation of the waves. If the initial tsunami profile has a wide range of frequency components, wave velocity varies slightly for these frequencies in deep water, causing shorter waves to arrive after a longer delay. The use of the dispersion term becomes essential to reproduce this effect. For far-field tsunamis that travel transoceanic distances, the Coriolis force due to earth's rotation should also be considered in the governing equations.

Initial conditions for tsunami simulations

For modeling of tsunami waves propagation, the initial disturbance of the water surface due to the amount of vertical slip is needed. JSCE usually determines the distribution of the vertical slip using the Mansinha and Smylie (1971) method, which assumes that elastic properties of the earth's crust near the fault are isotropic and homogeneous (i.e., Poisson Ratio v is 0.25 and Lame's constants μ and λ are equal). If the conditions of isotropy and homogeneity cannot be assumed, JSCE recommends using the Okada (1985) method, which is more general.

Fault motions that generate large tsunamis are assumed to last from several tens to approximately 120 seconds. Fault durations of this order are not significantly different than an instantaneous displacement of the seafloor in terms of the numerical simulation of the tsunami waves. Dynamics of the slip should be considered if $C.\ T_v/L$ is greater than 0.04, where C is the tsunami wave velocity, T_v is the duration of the fault motion, and L is the wavelength of the tsunami wave in the direction of the width of the fault plane.

The initial water-surface elevation is set equal to the vertical slip amount if fault motion is considered essentially instantaneous. The initial water-surface elevation is the still-water elevation if a dynamic vertical slip amount is used to characterize the tsunamigenic effect of a long- duration fault motion. In both cases, the depth- integrated initial flux is set to zero.

Boundary conditions for tsunami simulations

Three general boundary conditions are needed in tsunami wave-propagation simulations: (1) the offshore boundary condition, (2) the onshore boundary condition, and (3) the overflow boundary condition.

Offshore boundary conditions are specified at boundaries through which the tsunami waves pass unaltered. It is rather difficult to prescribe boundary conditions that eliminate all reflection of outgoing waves back into the computational domain (see Chapter 5 in Vreugdenhil 1994). The conventional method is to use discharge flux for a progressive wave at the boundary based on the method of characteristics (Aida 1969, 1970, 1974; Iwasaki and Yo 1974). The free-transmissionboundary condition can also be specified using a virtual complete reflecting wall at the open boundary (Hino and Nakaza 1988; Imamura 2001) with careful placement of the wall to achieve accuracy. The latter method for specification of an open-boundary condition is applicable only when the incidence angle of the waves on the boundary is nearly normal. Both boundary conditions are applicable for tsunami waves coming into the computational domain, a condition that is required for near-field computation of a tsunami generated in the far field.

Onshore boundary conditions are specified at the coastline. The nature of the boundary condition depends on whether runup on the land is considered. If tsunami runup is not considered, a complete reflection-boundary condition is specified by setting the discharge flux at the coastline to zero. This condition is applied only when water depth near the coastline is sufficient so that the tsunami rundown does not expose the bottom. If tsunami runup is considered, or if the water depth near the coastline is shallow, boundary conditions related to the runup front are used. The shore topography is approximated using a series of steps normal to the shoreline. At every time step of the numerical solution, the location of the tsunami front is determined based on computed water depth in computational grid cells. The method of Iwasaki and Mano (1979) is used in which (1) the tsunami front is assumed to be located at the boundary of the cell in which the computed sum of wave height and still-water depth is positive and the cell in which the computed sum is zero or negative, (2) the total water depth at the cell boundary for calculation of the discharge flux is taken as the sum of the still- water depth at the boundary and the higher wave height of the two neighboring cells, (3) the discharge flux is estimated assuming that the slope of the water surface can be approximated to the first order by the slope from the water level at the wave front to the bottom depth of the neighboring cell, and (4) when the total water depth approaches zero, the advection term is neglected.

Overflow boundary conditions are specified when the tsunami waves overtop a structure such as breakwaters, dikes, or revetments (Figure 6-2). If the structure is modeled as part of the ground topography, the onshore boundary conditions for runup computations are used. If the structure is located offshore, usually placed at the boundary of a grid cell in the discretization, two formulae are used depending on whether the structure is offshore or at the coastline.

For offshore structures, the formula by Honma (1940) is used to compute the discharge over the structure for complete and incomplete overflows:

$$q = \mu h_1 \sqrt{2 g h_1} \qquad \text{when } h_2 \le \frac{2}{3} h_1$$

$$q = 2.6\mu h_2 \sqrt{2 g(h_1 - h_2)} \qquad \text{when } h_2 > \frac{2}{3} h_1$$

where h_1 and h_2 are the water depths measured from the top of the structure in the front and the back of the structure, μ=0.35, and g is the acceleration due to gravity. When the waves do not flow over the structure, a no-flow boundary condition is imposed at the structure allowing complete reflection of the tsunami waves.

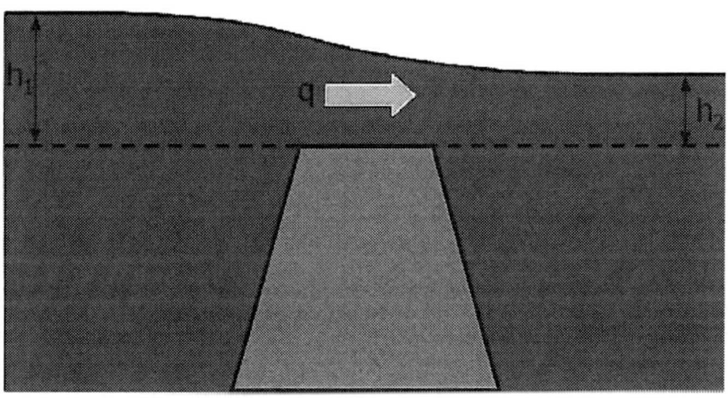

Figure 6-2. Schematic representation of tsunami wave overflow of offshore structures (JSCE, 2002), adapted with permission.

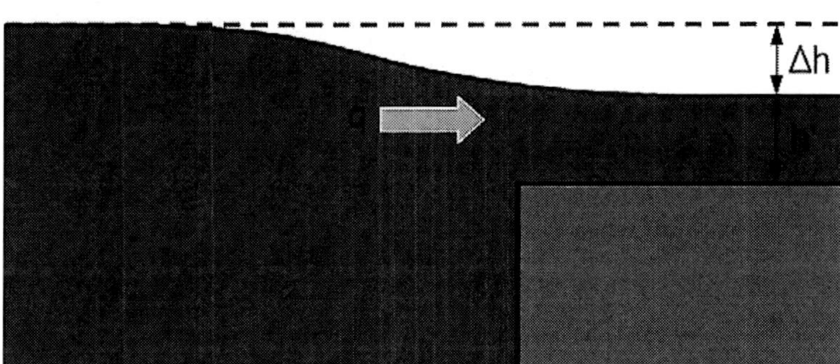

Figure 6-3. Schematic representation of tsunami overflow of coastal structures (JSCE, 2002), adapted with permission.

For structures that are located at the coastline, the discharge on the top of the structure is estimated using the broad-crested weir formula (Aida 1977):

$$q = 0.6\, h\sqrt{g\Delta h}$$

where h is the depth of water measured from the top of the structure and Δh is the drop in water level on the top of the structure from its original position (Figure 6-3).

Spatial grid considerations

JSCE (2002) described several considerations for choosing the size of the spatial grid for use in the numerical simulation of tsunamis, which are described as adequate to obtain sufficient accuracy when used with a staggered leapfrog scheme for discretization of governing differential equations:

1. In the tsunami source region, the grid size is determined based on dimensions of the source region and the spatial scale of the tsunami; as a rule of thumb, grid size equal to 1/20th of one wavelength of the tsunami is used.
2. In the tsunami propagation region, the grid size is determined based on refraction of tsunami waves caused by seafloor bathymetric features and the spatial scale of the tsunami; in regions of simple bathymetry, grid size 1/20th of one wavelength of the tsunami may suffice; however, in regions where refraction dominates, grid size as small as 1/100th of the tsunami wavelength may be required.
3. In shallow waters near the coast, the spatial grid size is determined based on spatial scale of the tsunami, slope of the seafloor, bathymetry and coastal topography, and size and shape of coastal structures. Coastal areas with simple bathymetry and topography and where effects of coastal structures are insignificant, grid sizes varying from approximately 328 ft (100 m) at a water depth of 164 ft (50 m) or less to approximately 82 ft (25 m) approaching the coastline may be sufficient. In the presence of ports and harbors, grid size less than 1/5th of the port entrance width may be necessary. Grid size less than 1/100th of the tsunami wavelength may be necessary in V-shaped bays where the ratio of the tsunami wavelength to the length of the bay is less than 6.

4. In the tsunami runup zone with simple beach topography, the following criterion is used:

$$\frac{\Delta x}{\alpha g T^2} \leq 7 \times 10^{-4}$$

when Manning's coefficient of roughness, n=0.03, is used, or

$$\frac{\Delta x}{\alpha g T^2} \leq 4 \times 10^{-4}$$

if the friction term is not considered in the governing equations, where Δx is the grid cell size, α is the slope of the beach, and T is the period of the tsunami waves.

Computational time step

The computational time step in the numerical scheme is determined from considerations of stability. Usually, the time steps are set to satisfy the Courant-Friedrichs-Levey (CFL) condition, which requires that the time step, Δt, should not be larger than the travel time of the wave over one grid cell:

$$\Delta t \leq \frac{\Delta x}{\sqrt{2 g h_{max}}}$$

where h_{max} is the maximum water depth. In a general case, where different grid-cell sizes are used in different regions of the computational domain, the time step is chosen such that the limit set by the CFL condition for the largest grid-cell size is not exceeded. In practice, however, the chosen time step for the tsunami simulation is smaller than the limit set by the CFL condition because of two reasons: to ensure that numerical errors are controlled, and to ensure that the time step is appropriate during high-speed current during the tsunami run-down when flow velocity may become greater than the wave celerity.

Bathymetric and topographic data

The latest available, high-resolution bathymetry data is recommended for use in the tsunami simulations. Generally, recent advances in echo- sounding techniques and their application to extensive areas provide high-quality bathymetry data. For far-field tsunami simulations, global- scale bathymetry datasets are required. The 5-minute and 2-minute gridded global-relief datasets are used.

Topographic data in the inundation area is prepared according to the latest available topographical maps. Numerical maps of the Geological Survey Institute and Japan Map Center are utilized. Changes are topography is also considered. Careful consideration of disparate georefrencing of global and Japanese datasets is also carried out to maintain consistency between datasets.

Application of the Japan approach to other locations

Preliminary results from the application of the Japan approach for estimation of tsunami-wave heights were described at the workshop, "Workshop on the Physics of Tsunami, Hazard Assessment Methods and Disaster Risk Management (Theories and Practices for Implementing Proactive Countermeasures)," held May 14-18, 2007 in Trieste, Italy. The workshop was jointly sponsored by the International Atomic Energy Agency (IAEA) and the Abdus Salam International Center for Theoretical Physics. Several IAEA member countries had expressed interest in 2006 to begin applying the JSCE approach to tsunami-hazard assessment at coastal nuclear power plant sites. The availability of the JSCE (2002) document has facilitated the implementation of the method in locations other than Japan.

Preliminary results from the application of the method to a coastal nuclear power plant site in India were presented. Two other locations, in Pakistan and in Egypt, were in early data-collection phase at the time of this writing. The conclusion drawn thus far from these early attempts to adopt the JSCE approach point to the reliance of the JSCE approach on the availability of a substantially large historical database of tsunami observations. These observations should contain tsunami-source characteristics (the JSCE approach only deals with earthquake sources) as well as tsunami runup information at or near the site of interest. At this time, few places outside Japan meet the data requirements needed to rigorously apply the JSCE approach. The IAEA member countries will continue to collect more data and attempt to adapt the JSCE approach to their locations of interest.

6.3. International Atomic Energy Agency

The International Atomic Energy Agency publishes a set of standards for safety evaluations of nuclear power plant sites. These standards are divided into Safety Fundamentals, Safety Requirements, and Safety Guides, supplemented by technical reports and preliminary technical documents, called TECDOCs, that contain illustrative examples.

The *Safety of Nuclear Installations,* Safety Series No. 110 (IAEA 1993), in the Safety Fundamentals series, presents basic concepts and principles underlying the regulation and safe management and operation of nuclear power installations. Chapter 5 of this document focuses on technical aspects of safety of nuclear installations, including those related to siting, design and construction, and operation and maintenance of the installation. *Site Evaluation for Nuclear Installations* (IAEA 2003a), in the Safety Requirements series, presents the requirements that are based on the concepts and principles documented in the fundamentals series and must be met to ensure safety of a nuclear installation. Chapter 3 of this document lists tsunamis as one cause of flooding, the evaluation of which is a requirement under the IAEA guidelines. *Flood Hazard for Nuclear Power Plants on Coastal and River Sites* (IAEA 2003b), in the Safety Guide series, provides recommendations to meet the requirements established in the Safety Requirements series (IAEA 2003a) with respect to flood hazards, including those caused by tsunamis, in the site evaluation for nuclear power plants that may be located on coastal and river sites.

IAEA (2003a) requirements state that the region in which a candidate site is located should be evaluated for potential for tsunamis. If such a potential exists, prehistorical and historical data should be collected. The frequency of occurrence, magnitude, and height of regional tsunamis should be estimated based on analysis of the collected prehistorical and historical data and comparison of the site region with similar regions that may have been studied in greater detail with regard to tsunamis. The hazards at the site should be determined based on physical and analytical modeling, where applicable. Coastal effects, such as the amplification of the waves due to shoaling and other coastal effects, should be considered.

IAEA (2003b) defines coastal sites as those that are located in open coastal regions and on enclosed or semi-enclosed bodies of water and whose hydrological responses cannot be compared to those of a small lake. Open coastal regions are defined as those areas of land that are directly exposed to a major body of water. Enclosed bodies of water includes lakes and reservoirs. Semi-enclosed bodies of water include lagoons, estuaries, gulfs, fjords, and rias.

IAEA (2003b) recommends that flooding resulting from a PMT be considered for a coastal site. It defines the PMT as "... the hypothetical tsunami having that combination of characteristics which will make it the most severe, in terms of flooding, that can reasonably be expected to occur at the site." Tsunamigenic sources that should be considered in site evaluation include earthquakes, landslides (including submarine landslides), submarine volcanoes, and falling ice. At a river site, floods resulting from large waves induced by volcanoes, landslides, and avalanches in water basins should be considered.

Preliminary investigation

Preliminary investigation related to flooding should be carried out at the site selection stage to establish the potential of flooding at the candidate site based on examination of available historical data and application of empirical and approximate methods to estimate the extreme flood (IAEA 2003b).

Catalogs of historical tsunamis should be searched to determine whether the site is subject to tsunamis. The potential for both near- and far- field sources to generate a tsunami that may affect the candidate site should be investigated. Nearby tide-gauge records may be used to estimate the maximum recorded tsunami wave heights. Based on coastal configuration, the correlation of tsunami effects at the site and that at the tide- gauge location may be established.

Preliminary estimates of the height and extent of the extreme wave should be made from analysis of collected historical data, bathymetry characteristics, and consideration of a conservative reference water level. Maps prepared for land-use planning and flood emergencies, aerial photographs and satellite imagery, and any available PMT studies in coastal areas may also be used for preliminary screening of the site.

If the preliminary investigation results in the conclusion that the site is potentially affected by tsunamis, a detailed study should be carried out to establish design bases for the proposed plant (IAEA 2003b).

Data collection

For a detailed tsunami-hazard assessment, a diverse set of data is needed. This data set includes hydrological, oceanographic and hydrographic, seismic, and geophysical data:

1. locations and hydrological characteristics of all nearby bodies of water, including streams, rivers, lakes, reservoirs, estuaries, gulfs and fjords, and oceans

2. hydrological and topographical characteristics of the site showing natural and artificial drainage features and any proposed changes
3. tides, sea-level anomalies, and water levels of nearby bodies of water
4. bathymetry of water bodies, particularly the near-shore region
5. locations and details of offshore structures, such as seawalls or breakwaters
6. near-shore currents induced by tide and wind
7. littoral drift
8. historical tsunami data, particularly tsunamis recorded at the site
9. seismic and geological data to determine source parameters
10. historical and potential future sites of subaerial and submarine landslides
11. volcanoes, including submarine-volcanic activity
12. sediment characteristics of the seafloor near proposed plant structures.

Detailed topographic and bathymetric maps are needed for tsunami-hazard assessment. If such data do not exist for the area near the plant site, surveys should be performed to prepare such maps or collect data from available survey methods. Detailed topographic data are needed in the tsunami-inundation zone. Detailed bathymetry data is needed from the shoreline out to a water depth of approximately 98-164 ft (30-50 m). Topographic and bathymetric maps or other sources of these data, such as digital elevation or depth data, should be combined at the shoreline to create a seamless map or dataset.

Simulation of tsunami waves

IAEA (2003b) recommends that the PMT generated from the worst-case tsunamigenicsource parameters should be determined. Both near- and far-field sources should be considered. Historical records of tsunamis, if available, should be used to validate the predictions of the numerical models used for simulation of the tsunami waves.

For submarine earthquake-generated tsunamis, the displacement of the sea floor should be estimated based on the assumption that the fault movement occurs in a semi-infinite elastic homogeneous medium. The fault movement is characterized by its location, depth, and length, and the width and dislocation of the fault plane. The initial tsunami waveform should be assumed to be the same as the static vertical displacement of the sea floor. Application of more sophisticated techniques that consider heterogeneity of fault movement should be carefully validated.

For landslide-generated tsunamis, dynamics of the mass movement should be determined to estimate the initial tsunami waveform. The parameters that describe the landslide dynamics include the velocity, the duration, and the discharge of the mass movement. For tsunamis generated by volcanic activity, IAEA (2003b) recommends that three generation mechanisms should be considered: the impact of rocks falling into the water after ejection, an underwater explosion that may result in a rapid rise of the water surface, and a caldera collapse. Characteristics of the volcanic mechanisms should be used to estimate the initial tsunami waveform. The simulation of the tsunami can be carried out in a manner similar to that for the earthquake-generated tsunami.

Far-field tsunamis may be treated as long linear waves while accounting for the Coriolis force during the propagation phase. The initial tsunami wavefield consists of many component frequencies that propagate at different celerities. For a tsunami that travels long distances, the difference in speed of component frequencies results in dispersion of the waves. IAEA (2003b) recommends a parameter P_a, which may be used to determine if dispersion effects should be considered in simulation of waves for a far-field tsunami:

$$P_a = \left(\frac{6\,h}{R}\right)^{1/3} \frac{a}{h}$$

where h is water depth, a is the horizontal dimension of the tsunamigenic source, and R is the distance from the tsunamigenic source to the site. If the value of P_a is larger than 4, frequency- dependent dispersion effects may be neglected and linear governing equations with accounting of the Coriolis force may be used. Otherwise, linearized Boussinesq equations that include the first-order effects of frequency-dependent dispersion and the effects of the Coriolis force should be used.

The long-wave approximation for near-field tsunamis may not be applicable since short-period components may also be significant near the tsunamigenic source. IAEA (2003b) recommends that simplifying assumptions for near-field tsunami should be carefully and critically examined and should be used only if they can be demonstrated to provide conservative results. IAEA (2003b) recommends that for a water depth exceeding 656 ft (200 m), linear long eave equations are applicable. For shallower regions, the shallow-water-wave theory with a term for bottom friction included in the governing equations should be used.

IAEA (2003b) recommends that near-shore modification of tsunami waves should be simulated using shallow-water equations, including the effects of bottom friction. Large oscillations of the water surface due to resonance of the tsunami waves with natural frequencies of the water body in the near-shore region should also be investigated. IAEA (2003b) recommends that tsunami runup be estimated from available approximate theories of the phenomena or from empirical relationships with careful consideration of the ranges of validity of the assumptions inherent in these approaches.

Tsunami impacts

Bottom shear caused by strong tsunami currents may be significant in shallow waters. The safety of plant structures may be affected by erosion and deposition of sediments by the tsunami currents, such as failure of a breakwater due to erosion or deposition of sediment at a safety-related cooling-water intake. IAEA (2003b) recommends that a dedicated analysis of these effects should be conducted. Measurement of sediment characteristics near the proposed locations of SSC important to safety may be required to make an assessment.

IAEA (2003b) recommends that three forms of tsunami waves should be considered in estimation of wave forces on structures:

1. non-breaking waves, which result in the tsunami acting as a rapidly rising tide
2. tsunami waves breaking far from the shoreline, which result in the tsunami waves developing into fully formed bores before reaching the shore
3. tsunami waves breaking near the shoreline, when the tsunami waves act as partially formed bores.

Both static and dynamic forces due to the three forms of the tsunami waves should be estimated on the structures of the nuclear power plant.

Combined events criteria

IAEA (2003b) recommends that appropriate combinations of extreme events with reference water levels (tides) and wind waves should be taken into consideration. Although precise estimation of the numerical probability that a given level of severity is exceeded by each separate event or by some combination of these events, conservative values of the probability that the given level of severity will be exceeded by each separate event and the likelihood that separate events may occur together in a combination should be estimated (IAEA 2003b). Further, reasonable values of the probability that a given level of severity may be exceeded by a combination of events should be estimated. Careful consideration of the duration of separate events in the combination is also required. In general, the greater the number of independent or partially dependent events in the combination, the lower the combined exceedance probability. IAEA (2003b) also notes that consideration of an excessive number of events in combination may result in overly conservative design bases and that considerable engineering judgement may be necessary in selecting appropriate combinations.

IAEA (2003b) recommends that tsunamis be considered in combination with waves caused by winds with a shorter recurrence interval. For determination of the runup, an ambient high-tide level should also be considered. IAEA (2003b) does not mention combining effects of seismic ground motion with the hazards from a concomitant tsunami.

REFERENCES

[1] 10 CFR Part 50. Code of Federal Regulations. Title 10, *Energy,* Part 50, "Domestic Licensing of Production and Utilization Facilities."

[2] 10 CFR Part 52. Code of Federal Regulations. Title 10, *Energy,* Part 52, "Early Site Permits; Standard Design Certifications; and Combined Licenses for Nuclear Power Plants."

[3] 10 CFR Part 100. Title 10, *Energy,* Part 100, "Reactor Site Criteria."

[4] Aida, I. (1969). "Numerical Experiments for the Tsunami Propagation - the 1964 Niigata Tsunami and the 1968 Tokachi-Oki Tsunami. " *Bulletin of the Earthquake Research Institute, University of Tokyo, 47*, 673-700.

[5] Aida, I. (1970). "A Numerical Experiment for the Tsunami Accompanying the Kanto Earthquake of 1923." *Bulletin of the Earthquake Research Institute, University of Tokyo, 48*, 73-86 (in Japanese).

[6] Aida, I. (1974). "Numerical Computation of a Tsunami Based on a Fault Origin Model of an Earthquake." *Earthquakes Series 2, 27*, 141-154 (in Japanese).

[7] Aida, I. (1977). "Numerical Experiments for Inundation of Tsunamis - Susaki and Usa, in the Kochi Prefecture." *Bulletin of the Earthquake Research Institute, University of Tokyo, 52*, 441460 (in Japanese).

[8] Aki, K. (1967). "Scaling Law of Seismic Spectrum." *Journal of Geophysical Research, 72*,1212-1231.

[9] Andrews, D. J. (1980). "A stochastic Fault Model 1. Static Case." *Journal of Geophysical Research, 85*, 3867-3877.

[10]Atwater, B. F. (1987). "Evidence for Great Holocene Earthquakes Along the Outer Coast of Washington State." *Science, 236*, 942-944.

[11]Atwater, B. (1996). "Coastal Evidence for Great Earthquakes in Western Washington." In: *Assessing Earthquake Hazards and Reducing Risk in the Pacific Northwest.* A.M. Rogers, T.J. Walsh, W.J. Kockelman, and G.R. Priest (eds.), *U.S. Geological Survey Professional Paper, 1560*, 77-90.

[12]Atwater, B. F. & Moore, A. L. (1992). "A Tunami about 1000 Years Ago in Puget Sound, Washington." *Science,* 258,1614-1617.

[13]Atwater, B. F., Satoko, M. R., Satake, K., Tsuji, Y., Ueda, K. & Yamaguchi, D. K. (2005). *The Orphan Tsunami of 1700: Japanese Clues to a Parent Earthquake in North America.* U.S. Geological Survey and University of Washington Press, Seattle, Washington.

[14]Beget, J. E. (2000). "Volcanic Tsunamis " In: *Encyclopedia of Volcanoes,* H. Sigurdsson (ed.), Academic Press, San Diego, California, pp. 1005-1013.

[15]Benson, W. E., Grimm, K. A. & Clague. J. J. (1997). "Tsunami Deposits Beneath Tidal Marshes on Northwestern Vancouver Island, British Columbia." *Quaternary Research, 48*, 192-204.

[16]Berge, C., Gariel, J. C. & Bernard, P. (1998). "A Very Broad-Band Stochastic Source Model Used for Near Source Strong Motion Prediction." *Geophysical Research Letters, 25*, 1063-1066.

[17]Bilek, S. L. & Lay, T. (1999). "Rigidity Variations with Depth Along Interplate Megathrust Faults in Subduction Zones." *Nature, 400*, 443-446.

[18]Bilek, S. L. & Lay, T. (2000). "Depth-Dependent Rupture Properties in Circum-Pacific Subduction Zones." In: *Geo-Complexity and the Physics of Earthquakes,* J.B. Rundle, D.L. Turcotte, and W. Klein (eds.), Geophysical Monograph, American Geophysical Union, Washington, D.C., pp. 165186.

[19]Bird, P. & Kagan, Y. Y. (2004). "Plate-Tectonic Analysis of Shallow Seismicity: Apparent Boundary Width, Beta-Value, Corner Magnitude, Coupled Lithosphere Thickness, and Coupling in 7 Tectonic Settings." *Bulletin of the Seismological Society of America, 94*, 2380-2399.

[20]Bondevik, S., Svendsen, J. L. & Mangerud, J. (1997). "Tsunami Sedimentary Facies Deposited by the Storegga Tsunami in Shallow Marine Basins and Coastal Lakes, Western Norway." *Sedimentology, 44*, 1115-1131.

[21]Booth, J. S., O'Leary, D. W., Popenoe, P. & Danforth, W. W. (1993). "U.S. Atlantic Continental Slope Landslides: Their Distribution, General Attributes, and Implications." In *Submarine Landslides: Selected Studies*

in the U.S. Exclusive Economic Zone, W.C. Schwab, H.J. Lee, and D.C. Twichell (eds.), U.S. Geological Survey Bulletin 2002, 14-22.

[22]Bouma, A. H. (1962). *Sedimentology of Some Flysch Deposits: A Graphic Approach to Facies Interpretation.* Elsevier, Amsterdam, the Netherlands.

[23]Bryn, P., Solheim, A., Berg, K., Lien, R., Forsberg, C. F., Haflidason, H. Ottesen, D. & Rise. L. (2003). "The Storegga Slide Complex: Repeated Large-Scale Sliding in Response to Climate Cyclicity." In *Submarine Mass Movements and Their Consequences,* J. Locat and J. Mienert (eds.), Kluwer Academic Publishers, Dordrecht, the Netherlands, 215-222.

[24]Bulat, J. (2003). "Imaging the Afen Slide from Commercial 3D Seismic Methodology and Comparisons with High-Resolution Data." In: *Submarine Mass Movements and Their Consequences,* J. Locat and J. Mienert (eds.), Kluwer Academic Publishers, Dordrecht, the Netherlands, 205-213.

[25]Burroughs, S. M. & Tebbens, S. F. (2005). "Power-Law Scaling and Probabilistic Forecasting of Tsunami Runup Heights." *Pure and Applied Geophysics, 162*, 331-342.

[26]Cannon, E. C., Biirgmann, R. & Owen, S. E. (2001). "Shallow Normal Faulting and Block Rotation Associated with the 1975 Kalapana Earthquake, Kilauea Volcano, Hawaii." *Bulletin of the Seismological Society of America, 91*, 1553-1562.

[27]Carracedo, J. C., Day, S. J., Guillouc, H. & Perez Torradod, F. J. (1999). "Giant Quaternary Landslides in the Evolution of La Palma and El Hierro, Canary Islands." *Journal of Volcanology and Geothermal Research, 94*, (1-4):169-190.

[28]Chanson, H. (2005). "Physical Modelling of the Flow Field in an Undular Tidal Bore." *Journal of Hydraulic Research, 3(3)*, 234-244.

[29]Chopra, A. K. (1995). *Dynamics of Structures: Theory and Applications to Earthquake Engineering,* Second Edition, Prentice-Hall, New Jersey.

[30]Dawson, A. G., Long, D. & Smith, D. E. (1988). "The Storegga Slides: Evidence from Eastern Scotland for a Possible Tunami." *Marine Geology, 82*, 271-276.

[31]Day, S. J., Heleno da Silvab, S. I. N. & Fonsecab, J. F. B. D. (1999). "A Past Giant Lateral Collapse and Present-Day Flank Instability of Fogo, Cape Verde Islands." *Journal of Volcanology and Geothermal Research, 94(1-4)*, 191-218.

[32]Dmowska, R. & Kostrov, B. V. (1973). "A Shearing Crack in a Semi-Space Under Plane Strain Conditions." *Archives of Mechanics, 25*, 421-440.

[33]Dragovich, J. D., Pringle, P. T. & Walsh, T. J. (1994). "Extent and Geometry of the Mid-Holocene Osceola Mudflow in the Puget Lowland — Implications for Holocene Sedimentation and Paleogeography." *Washington Geology, 22(3)*, 326.

[34]Elsworth, D. & Day, S. J. (1999). "Flank Collapse Triggered by Intrusion: the Canarian and Cape Verde Archipelagoes." *Journal of Volcanology and Geothermal Research, 94(1-4)*, 323-340.

[35]Federal Emergency Management Agency (FEMA). (2005). *Coastal Construction Manual: Principles and Practices of Planning, Siting, Designing, Constructing, and Maintaining Residential Buildings in Coastal Areas,* FEMA 55, Washington, D.C.

[36]Fine, I. V., Rabinovich, A. B., Kulikov, E. A., Thompson, R. E. & Bonhold, B. D. (1999). "Numerical Modeling of Landslide-Generated Tsunamis with Application to the Skagway Harbor Tsunami of November 3, 1994." *Proceedings of International Conference on Tsunamis,* Paris, France — 1998, pp. 211-214.

[37]Fine, I. V., Rabinovich, A. B., Bornhold, B. D., Thomson, R. E. & Kulikov, E. A. (2005). "The Grand Banks Landslide-Generated Tsunami of November 18, 1929: Preliminary Analysis and Numerical Modeling." *Marine Geology,* 215, 45-57.

[38]Fritz, H. M., Hager, W. H. & Minor, H. E. (2001). "Lituya Bay Case: Rockslide Impact and Wave Run-Up." *Science of Tsunami Hazards, 19(1)*, 322.

[39]Fritz, H. M., Hager, W. H. & Minor, H. E. (2003a). "Landslide-Generated Impulse Waves, Part 1: Instantaneous Flow Fields." *Experiments in Fluids, 35*, 505-519.

[40]Fritz, H. M., Hager, W. H. & Minor, H. E. (2003b). Landslide-Generated Impulse Waves, Part 2: Hydrodynamic Impact Craters." *Experiments in Fluids, 35*, 520-532.

[41]Fritz, H. M., Hager, W. H. & Minor, H. E. (2004). "Near-Field Characteristics of Landslide- Generated Impulse Waves." *Journal of Waterway, Port, Coastal, and Ocean Engineering, 130*, 287302.

[42]Fryer, G. J., Watts, P. & Pratson, L. F. (2004). "Source of the Great Tsunami of 1 April 1946: a Landslide in the Upper Aleutian Forearc." *Marine Geology, 203*, 201-218.

[43]Gardner, J. V. & Mayer, L. A. (1998). *Cruise Report RV Ocean Alert Cruse A2-98-SC• Mapping the Southern California Continental Margin.* U.S. Geological Survey Open-File Report 98-475.

[44]Gardner, J. V., Hughes Clarke, J. E. & Mayer, L. A. (1999). *Cruise*

Report: RV Coastal Surveyor Cruise C-1-99-SC: Multibeam Mapping of the Long Beach, California, Continental Shell U.S. Geological Survey Open-File Report 99-360.

[45]Gardner, J. V., van den Ameele, E. J. & Dartnell, P. (2001). *Multibeam Mapping of the Major Deltas of Southern Puget Sound, Washington.* U.S. Geological Survey Open-File Report 2001266, URL: http://pubs.er.usgs.gov/usgspubs/ofr/ofr01266. Accessed March 12, 2008.

[46]Geist, E. L. (2005). *Local Tsunami Hazards in the Pacific Northwest from Cascadia Subduction Zone Earthquakes.* Professional Paper 1661-B, U.S. Geological Survey, Reston, Virginia.

[47]Geist, E. L. & Dmowska, R. (1999). "Local Tsunamis and Distributed Slip at the Source." *Pure and Applied Geophysics, 154*, 485-512.

[48]Geist, E. L. & Parsons, T. (2006). "Probabilistic Analysis of Tsunami Hazards." *Natural Hazards, 37*, 277-314.

[49]Geller, R. J. (1976). "Scaling Relations for Earthquake Rource Parameters and Magnitudes." *Bulletin of the Seismological Society of America, 66*, 1501-1523.

[50]Gica, E., Spillane, M., Titov, V., Chamberlin, C. & Newman, J. C. (2008). *Development of the Forecasting Propagation Database for NOAA 's Short-term Inundation Forecast for Tsunamis (SIFT).* NOAA Technical Memorandum OAR PMEL-139, Pacific Marine Environmental Laboratory, Seattle, Washington.

[51]Gonzalez, F. I., Sherrod, B. L., Atwater, B. F., Frankel, A. P., Palmer, S. P., Holmes, M. L., Karlin, R. E., Jaffe, B. E., Titov, V. V., Mofield, H. O. & Venturato, A. J. (2003). "Puget Sound Tsunami Sources-2002 Workshop Report." In *A Contribution to the Inundation Mapping Project of the U.S. National Tsunami Hazard Mitigation Program,* NOAA OAR Special Report, NOAA/OAR/PMEL.

[52]Gonzalez, F. I, Geist, E., Synolakis, C., Arcas, D., Bellomo, D., Carlton, D., Horning, T., Jaffe, B., Johnson, J., Kanoglu, U., Mofjeld, H., Newman, J., Parsons, T., Peters, R., Peterson, C., Priest, G., Titov, V., Venturato, A., Weber, J., Wong, F. & Yalciner. (2006). *Seaside, Oregon Tsunami Pilot Study–Modernization of FEMA Flood Hazard Maps,* NOAA OAR Special Report, NOAA/OAR/PMEL, Seattle, Washington.

[53]Gonzalez, F. I., Bernard, E., Dunbar, P., Geist, Jaffe, E., Kanoglu, U., Locat, J., Mofjeld, H., Moore, A., Synolakis, C. & Titov, V. (2007). *Scientific and Technical Issues in Tsunami Hazard Assessment of Nuclear Power Plant Sites,* NOAA Technical Memorandum OAR PMEL-136, Pacific Marine Environmental Laboratory, Seattle, Washington.

[54]Gray, J. P. & Monaghan, J. J. (2003). "Caldera Collapse and the Generation of Waves." *Geochemistry, Geophysics, Geosystems, 4(2)*, 1015. doi: 10.1029/2002G0000411.

[55]Greene, H. G. & Ward, S. N. (2003). "Mass Movement Features Along the Central California Margin and Their Modeled Consequences for Tsunami Generation." In *Submarine Mass Movements and Their Consequences,* J. Locat and J. Mienert (eds.), Kluwer Academic Publishers, Dordrecht, the Netherlands, 343-356.

[56]Hampton, M. A., Lee, H. J. & Locat, J. (1996). "Submarine Landslides." *Reviews in Geophysics, 34(1)*, 33-59.

[57]Heinrich, P. (1992). "Nonlinear Water Waves Generated by Submarine and Aerial Landslides." *Journal of Waterway, Port, Coastal, and Ocean Engineering, 118*, 249-266.

[58]Heinrich, P., Mangeney, A., Guibourg, S., Roche, R., Boudon, G. & Chemin, J. L. (1998). "Simulation of Water Waves Generated by a Potential Debris Avalanche in Montserrat, Lesser Antilles." *Geophysical Research Letters, 25(19)*, 3697-3700.

[59]Heinrich, P., Piatanesi, A. & Hebert, H. (2001). "Numerical Modelling of Tsunami Generation and Propagation from Submarine Slumps: The 1998 Papua New Guinea Event. " *Geophysical Journal International, 145*, 97-111.

[60]Hendron, A. J. & Patten, F. D. (1985). *The Vaiont Slide.* U.S. Army Corps of Engineers Technical Report GL-85-8.

[61]Herrero, A. & Bernard, P. (1994). "A Kinematic Self-Similar Rupture Process for Earthquakes." *Bulletin of the Seismological Society of America, 84*, 1216-1228.

[62]Hino, M. & Nakaza, E. (1988). "Application of New Non-reflection Boundary Scheme to Plane Two-dimensional Problems in Numerical Wave Motion Analysis." In: *Proceedings of the 35th Conference on Coastal Engineering, Japan Society of Civil Engineers,* pp. 262-266 (in Japanese).

[63]Honma, H. (1940). "Flow Rate Coefficient of Low- overflow Dams." *Civil Engineering, 26(6)*, 635645, 26(9):849-862 (in Japanese).

[64]Huggel, C., Haeberli, W., Kaab, A., Bieri, D. & Richardson, S. D. (2004). "An Assessment Procedure for Glacial Hazards in the Swiss Alps." *Canadian Geotechnical Journal, 41*,1068-1083.

[65]Hughes-Clarke, J. E., Mayer, L. A. & Wells, D. E. (1996). "Shallow-Water Imaging Multibeam Sonars: A New Tool for the Investigating Seafloor Processes in the Coastal Zone and on the Continental Shelf."

Marine Geophysical Letters, 18, 607-629.

[66]Ichinose, G. A., Anderson, J. G., Satake, K., Schweickert, R. A. & Lahern, M. M. (2000). "The Potential Hazard from Tsunami and Seiche Waves Generated by Large Earthquakes within Lake Tahoe, California-Nevada." *Geophysical Research Letters, 27*, 1203-1206.

[67]Imamura, F. (2001). "Numerical Analysis of 1771 Meiwa Big Tsunami and Tsunami Stone Dislocation in Ishigaki Island." *Technical Report of Tsunami Engineering,* 18, 61-72 (in Japanese).

[68]Imamura, F., Yalciner, A. C. & Ozyurt, G. (2006). *Tsunami Modelling Manual.* April 2006 Draft, Tohoku University, Sendai Japan and Middle East Technical University, Ankara, Turkey.

[69]Imran, J., Parker, G., Locat, J. & Lee. H. (2001). "A 1-D Numerical Model of Muddy Subaqueous and Subaerial Debris Flows." *Marine and Petroleum Geology, 22*,187-194.

[70]International Atomic Energy Agency (IAEA). (1993). *The Safety of Nuclear Installations.* Safety Series No. 110, IAEA, Vienna, Austria.

[71]International Atomic Energy Agency (IAEA). (2003a). *Site Evaluation for Nuclear Installations: Safety Requirements.* No. NS-R-3, IAEA, Vienna, Austria.

[72]International Atomic Energy Agency (IAEA). (2003b). *Flood Hazard for Nuclear Power Plants on Coastal and River Sites* Safety Guide No. NS-G-3.5, IAEA, Vienna, Austria.

[73]Issler, D., De Blasio, F. V., Elverhoi, A., Byrn, P. & Lien, R. (2005). "Scaling Behaviour of Clay- Rich Submarine Debris Flows. " *Marine and Petroleum Geology, 22*, 187-194.

[74]Iwasaki, T. & Yo, T. (1974). "Numerical Experiments of Sanriku Big Tsunami." *In: Proceedings of the 21st Conference on Coastal Engineering,* Japan Society of Civil Engineers, pp. 83-89 (in Japanese).

[75]Iwasaki, T. & Mano, A. (1979). "Two- Dimensional Numerical Computation of Tsunami Run-Ups in the Eulerian Description." In: *Proceedings of the 26th Conference on Coastal Engineering,* Japan Society of Civil Engineers, pp. 70-74 (in Japanese).

[76]Japan Society of Civil Engineers (JSCE). (2002). *Tsunami Assessment Method for Nuclear Power Plants in Japan* (English translation dated February 2002). The Tsunami Evaluation Subcommittee of The Nuclear Civil Engineering Committee, Tokyo, Japan.

[77]Jeyakumaran, M. & Keer, L. M. (1994). "Curved Slip Zones in an Elastic Half-Plane." *Bulletin of the Seismological Society of America, 84*,1903-1915.

[78]Jeyakumaran, M., Rudnicki, J. W. & Keer, L. M. (1992). "Modeling Slip Zones with Triangular Dislocation Elements." *Bulletin of the Seismological Society of America, 82*, 2153-2169.

[79]Jiang, L. & LeBlond, P. H. (1992). "The Coupling of a Submarine Slide and the Surface Waves Which it Generates." *Journal of Geophysical Research, 97*, 12,731-12,744.

[80]Jiang, L. & LeBlond, P. H. (1993). "Numerical Modeling of an Underwater Bingham Plastic Mudslide and the Waves Which it Generates." *Journal of Geophysical Research, 98*, 10,30310,317.

[81]Jiang, L. & LeBlond, P. H. (1994). "Three- Dimensional Modeling of Tsunami Generation due to a Submarine Mudslide." *Journal of Physical Oceanography, 24*, 559-572.

[82]Johnson, J. M. & Satake, K. (1997). "Estimation of Seismic Moment and Slip Distribution of the April 1, 1946, Aleutian Tsunami Earthquake." *Journal of Geophysical Research, 102*(B6), 11,76511,774.

[83]Kagan, Y. Y. (2002). "Seismic Moment Distribution Revisited: I. Statistical Results." *Geophysical Journal International, 148*, 520-541.

[84]Kagan, Y. Y. & Jackson, D. D. (1991). "Seismic Gap Hypothesis: Ten Years After." *Journal of Geophysical Research, 96*, 21,419-21,431.

[85]Kagan, Y. Y. & Jackson, D. D. (1995). "New Seismic Gap Hypothesis: Five Years After." *Journal of Geophysical Research, 100*, 3943-3959.

[86]Kanamori, H. (1985). "Non-Double-Couple Seismic Source." *Proceedings of the 23r[d] General Assembly of the International Association of Seismology and Physics of the Earth's Interior,* p. 425, Tokyo, Japan.

[87]Kanamori, H. & Anderson, D. L. (1975). "Theoretical Basis of Some Empirical Relations in Seismology." *Bulletin of the Seismological Society of America, 65*, 1073-1095.

[88]Kanamori, H., Given, J. W. & Lay, T. (1984). "Analysis of Seismic Body Waves Excited by the Mount St. Helens Eruption of May 18, 1980." *Journal of Geophysical Research, 89*, 1856-1866.

[89]Kanamori, H., EktrOm, G., Dziewonski, A. M., Barker, J. S. & Sipkin, S. A. (1993). "Seismic Radiation by Magma Injection: An Anomalous Seismic Event Near Tori Shima, Japan." *Journal of Geophysical Research, 98*, 6511-6522.

[90]Kanoglu U. & Synolakis, C. E. (2006). "Initial Value Problem Solution of Nonlinear Shallow Water-Wave Equations." *Physical Review Letters, 97*, 148501.

[91]Kiersch, G. A. (1964). "Vaiont Reservoir Disaster." *Civil Engineering, 34*, 32-39.

[92]Kirby, S., Geist, E., Lee, W. H. K., Scholl, D. & Blakely, R. (2006). *Tsunami Source Characterization for Western Pacific Subduction Zone: A Preliminary Report.* U.S. Geological Survey (USGS) Tsunami Subduction Zone Working Group, USGS white paper.

[93]Klein, F. W., Frankel, A. D., Mueller, C. S., Wesson, R. L. & Okubo, P. G. (2001). "Seismic Hazard in Hawaii: High Rate of Large Earthquake and Probabilistic Ground-Motion Maps." *Bulletin of the Seismological Society of America, 91*, 479-498.

[94]Koch C. & Chanson, H. (2005). *An Experimental Study of Tidal Bores and Positive Surges: Hydrodynamics and Turbulence of the Bore Front.* Department of Civil Engineering Report CH56/05, The University of Queensland, Brisbane, Australia.

[95]Latter J. H. (1981). "Tsunamis of Volcanic Origin: Summary of Causes, with Particular Reference to Krakatau, 1883." *Bulletin of Volcanology, 44*, 467-490.

[96]Lavallee D., Liu, P. & Archulet, R. J. (2006). "Stochastic Model of Heterogeneity in Earthquake Slip Spatial Distributions." *Geophysical Journal International, 165*, 622-640.

[97]Lay T. & Wallace, T. C. (1995). *Modern Global Seismology.* International Geophysics Series, 58, Academic Press, San Diego, California.

[98]Lee, H. J., Syvitski, J. P. M., Parker, G., Orange, D., Locat, J., Hutton, E. W. H. & Imran, J. (2002). "Distinguishing Sediment Waves from Slope Failure Deposits: Field Examples, Including the `Humboldt Slide' and Modelling Results." *Marine Geology, 192*, 79-104.

[99]Lee H., Kayen, R. E., Gardner, J. V. & Locat, J. (2003). "Characteristics of Several Tsunamigenic Landslides." In: *Submarine Mass Movements and Their Consequences,* J. Locat and J. Meinert (eds.), Kluwer Academic Publishers, Dordrecht, The Netherlands, pp. 357-366.

[100] Lee, H. J., Normark, W. R., Fisher, M. A., Greene, G., Edwards, B. D. & Locat, J. (2004). "Timing and Extent of Submarine Landslides in Southern California." *Offshore Technology Conference,* Houston, Texas, Paper 16744.

[101] Legros, F. & Druitt, T. H. (2000). "On the Implacement of Ignimbrite in Shallow Water Environments." *Journal of Volcanology and Geothermal Research, 95*, 9-22.

[102] Lipman, P. W. (1995). "Declining Growth of Mauna Loa During the Last 100,000 Years: Rates of Lava Accumulation vs. Gravitational Subsidence." In: *Mauna Loa Revealed: Structure, Composition, History, and Hazards,* J.M. Rhodes and J.P. Lockwood (eds.) American

Geophysical Union, Washington, D.C. Geophysical Monograph 92:45-80.

[103] Lipman P.W., W.R. Normark, J.G. Moore, J.B. Wilson, and C.E. Gutmacher. 1988. "The Giant Submarine Alika Debris Slide, Mauna Loa, Hawaii." *Journal of Geophysical Research, 93*(B5), 4279-4299.

[104] Lipman, P. W., Sisson, T. W., Ui, T., Naka, J. & Smith, J. R. (2002). "Ancestral Submarine Growth of Kilauea Volcano and Instability of its South Flank." In: *Hawaiian Volcanoes: Deep Underwater Perspectives,* E. Takahashi, P.W. Lipman, M.O. Garcia, J. Naka, and S. Aramaki (eds.), Geophysical Monograph 128, American Geophysical Union, pp. 161-191.

[105] Lipman, P. W., Eakins, B. W. & Yokose, H. (2003). "Ups and Downs on Spreading Flanks of Ocean- Island Volcanoes: Evidence from Mauna Loa and Kilauea." *Geology, 31*, 841-844.

[106] Locat, J. & Lee, H. J. (2002). "Submarine Landslides: Advances and Challenges." *Canadian Geotechnical Journal, 39*, 193-212.

[107] Locat J., Lee, H. J., Locat, P. & Imran, J. (2004). "Numerical Analysis of the Mobility of the Palos Verdes Debris Avalanche, California, and its Implications for the Generation of Tsunamis." *Marine Geology, 203*, 269-280.

[108] Lockridge, P. A., Whiteside, L. S. & Lander, J. F. (2002). "Tsunamis and Tsunami-Like Waves of the Eastern United States." *Science of Tsunami Hazards, 20(3)*, 120-157.

[109] Lopez A. M. & Okal, E. A. (2006). "A Seismological Reassessment of the Source of the 1946 Aleutian 'Tsunami' Earthquake." *Geophysical Journal International, 165*, 835-849.

[110] Lynett, P. & Liu, P. L. F. (2002). "A Numerical Study of Submarine Landslide-Generated Waves and Runup." *Proceedings of the Royal Society of London A, 458*, 2885-2910.

[111] Lipman, P. W., Eakins, B. W. & Yokose, H. (2003). "Ups and Downs on Spreading Flanks of Ocean- Island Volcanoes: Evidence from Mauna Loa and Kilauea." *Geology, 31*, 841-844.

[112] Lynett P. & Liu, P. L. F. (2006). "Three- Dimensional Runup due to Submerged and Subaerial Landslides." In: *Caribbean Tsunami Hazard,* A. Mercado and P.L.F. Liu (eds.). World Scientific Publishing Company, Singapore, pp. 289-307.

[113] Ma K. F., Kanamori, H. & Satake, K. (1999). "Mechanism of the 1975 Kalapana, Hawaii, Earthquake Inferred from Tsunami Data." *Journal of Geophysical Research, 104*(B6), 13,153-13,167.

[114] Mader, C. L. (2001). "Modelling the La Palma Landslide Tsunami." *Science of Tsunami Hazards, 19(3)*, 150-170.

[115] Mader C. L. (2004). *Numerical Modeling of Water Waves.* Second Edition, CRC Press, Boca Raton, Florida.

[116] Mader, C. L. & Gittings, M. L. (2002). "Modeling the 1958 Lituya Bay Mega-Tsunami, II." *Science of Tsunami Hazards, 20*, 241-250.

[117] Mai, P. M. & Beroza, G. C. (2002). "A Spatial Random Field Model to Characterize Complexity in Earthquake Slip." *Journal of Geophysical Research, 107*(B11)L2308. doi:10.1029/2001JB000588.

[118] Mansinha, L. & Smylie, D. E. (1971). "The Displacement Field of Inclined Faults." *Bulletin of the Seismological Society of America, 61(5)*, 1433-1440.

[119] Marks, K. M. & Smith, W. H. F. (2006). "An Evaluation of Publicly Available Global Bathymetry Grids." *Marine Geophysical Researches* 27:19-34.

[120] Masson D. G., Watts, A. B., Gee, M. J. R., Urgeles, R., Mitchell, N. C., Le Bas, T. P. & Canals. M. (2002). "Slope Failure on the Flanks of the Western Canary Islands." *Earth-Science Reviews, 57, 1-35.*

[121] Mastronuzzi G. & Sanso, P. (2000). "Boulders Transport by Catastrophic Waves Along the Ionian Coast of Apulia (Southern Italy)." *Marine Geology, 170*, 93-103.

[122] Mayer L. A. (2006). "Frontiers in Seafloor Mapping and Visualization." *Marine Geophysical Researches, 27*, 7-17.

[123] McAdoo, B. G., Pratson, L. F. & Orange. D. L. (2000). "Submarine Landslide Geomorphology, U.S. Continental Slope." *Marine Geology, 169*, 103-136.

[124] McGregor, B. A., Rothwell, R. G., Kenyon, N. H. & Twichell, D. C. (1993). "Salt Tectonics and Slope Failure in an Area of Salt Domes in the Northwestern Gulf of Mexico." In: *Submarine Landslides: Selected Studies in the U.S. Exclusive Economic Zone,* W.C. Schwab, H.J. Lee, and D.C. Twichell (eds.), U.S. Geological Survey Bulletin 2002, pp. 92-96.

[125] Miller, D. J. (1960). "Giant Waves in Lituya Bay, Alaska." In: *Shorter Contributions to General Geology.* U.S. Geological Survey Professional Paper 354-C:51-86.

[126] Moore, A. L. (2000). "Landward Fining in Onshore Gravel as Evidence for a Late Pleistocene Tsunami on Molokai, Hawaii." *Geology, 28(3)*, 247-250.

[127] Moore, J. G. & Clague, D. A. (1992). "Volcano Growth and Evolution of the Island of Hawaii." *Geologic Society of America Bulletin, 104*, 1471-1484.

[128] Moore, J. G., Clague, D. A., Holcomb, R. T., Lipman, P. W., Normark,

W. R. & Torresan, M. E. (1989). "Prodigious Submarine Landslides on the Hawaiian Ridge. *Journal of Geophysical Research, 94*, 17,465-17,484.

[129] Moore, J. G., Normark, W. R. & Holcomb, R. T. (1994a). "Giant Hawaiian Landslides." *Annual Review of Earth and Planetary Sciences, 22*, 119144.

[130] Moore, J. G., Normark, W. R. & Holcomb, R. T. (1994b). "Giant Hawaiian Underwater Landslides." *Science, 264*, 46-47.

[131] Morgan, J. K., Moore, G. F. & Clague, D. A. (2003). "Slope Failure and Volcanic Spreading Along the Submarine South Flank of Kilauea Volcano, Hawaii." *Journal of Geophysical Research, 108*(B9), 2415. doi:10.1029/2003JB002411.

[132] Murty, T. S. (1979). "Submarine Slide-Generated Water Waves in Kitimat Inlet, British Columbia." *Journal of Geophysical Research, 84*(C12), 7777-7779.

[133] Nanayama, F., Satake, K., Furukawa, R., Shimokawa, K., Atwater, B. F., Shigeno, K. & Yamaki, S. (2003). "Unusually Large Earthquakes Inferred from Tsunami Deposits Along the Kuril Trench." *Nature, 424(6949)*, 660-663.

[134] National Geophysical Data Center (NGDC). (2007). NOAA/WDC Historical Tsunami Database at NGDC, URL: http://www.ngdc.noaa.gov/hazarditsu_db.shtml. Accessed August 25, 2008.

[135] National Science and Technology Council (NSTC). (2005). *Tsunami Risk Reduction for the United States: A Framework for Action.* NSTC, Executive Office of the President of the United States, Washington, D.C.

[136] Nuclear Power Corporation of India Limited (NPCIL). (2007). Press Release, URL: http://www.npcil.nic.in/riews 20Jan05 01.asp. Accessed August 25, 2008.

[137] Newhall, C. G. & Self, S. (1982). "The Volcanic Explosivity Index (VEI): An Estimate of Explosive Magnitude for Historical Volcanism." *Journal of Geophysical Research, 87*, 1231-1238.

[138] Nomanbhoy N. & Satake, K. (1995). "Generation Mechanism of Tsunamis from the 1883 Krakatau Eruption." *Geophysical Research Letters, 22*, 509512.

[139] Okada, U. (1985). "Surface Deformation due to Shear and Tensile Faults in a Half-Space." *Bulletin of the Seismological Society of America, 75(4)*, 1135-1154.

[140] Okal, E. A. (1988). "Seismic Parameters Controlling Far-Field Tsunami Amplitudes: A Review." *Natural Hazards, 1, 67-96.*

[141] Okal, E. A., Borrero, J. C. & Synolakis, C. E. (2006). "Evaluation of Tsunami Risk from Regional Earthquakes at Pisco, Peru. " *Bulletin of Seismological Society of America, 96*, 1634-1648.

[142] O'Leary, D. W., Hughes-Clarke, J., Dobson, M. & Williams, S. R. J. (1987). "GLORIA Sonar Survey of the New England-Nova Scotian Continental Margin: The Rise Revealed." *Geotimes,* March 1988, pp. 22-24.

[143] O'Leary, D. W. (1993). "Submarine Mass Movement, a Formative Process of Passive Continental Margins: The Munson-Nygren Landslide Complex and the Southeast New England Landslide Complex." In: *Submarine Landslides: Selected Studies in the Exclusive Economic Zone,* W.C. Schwab, H.J. Lee, and D.C. Twichell (eds.), U.S. Geological Survey Bulletin 2002, pp. 23-39.

[144] Pararas-Carayannis, G. (2002). "Evaluation of the Threat of Mega-Tsunami Generation from Postulated Massive Slope Failures of Island Stratovolcanoes on La Palma, Canary Islands and on the Island of Hawaii." *Science of Tsunami Hazards, 20(5)*, 251-277.

[145] Peters, B., Jaffe, B., Gelfenbaum, G. & Peterson, C. (2003). *Cascadia Tsunami Deposit Database.* U.S. Geological Survey Open-File Report 03-13, and electronic database and GIS coverage [URL: http://geopubs.wr.usgs.gov/openfile/of03-13/]. Accessed March 12, 2008.

[146] Pinegina, T. K. & Bourgeois, J. (2001). "Historical and Paleo-Tsunami Deposits on Kamchatka, Russia: Long-Term Chronologies and Long-Distance Correlations." *Natural Hazards and Earth Systems Sciences, 1, 177-185.*

[147] Pinegina, T. K., Bourgeois, J., Bazanova, L. I., Melekestsev, I. V. & Braitseva, O. A. (2003). "A Millennial-Scale Record of Holocene Tsunamis on the Kronotskiy Bay Coast, Kamchatka, Russia." *Quaternary Research, 59*, 36-47.

[148] Popenoe, P., Schmuck, E. A. & Dillon, W. P. (1993). "The Cape Fear Landslide: Slope Failure Associated with Salt Diapirism and Gas Hydrate Decomposition." In: *Submarine Landslides: Selected Studies in the U.S. Exclusive Economic Zone,* W.C. Schwab, H.J. Lee, and D.C. Twichell (eds.), U.S. Geological Survey Bulletin 2002, pp. 40-53.

[149] Ranalli, G. (2001). *Rheology of the Earth.* Second Edition, Chapman and Hall, Dordrecht, The Netherlands.

[150] Reyes, L. A. (2006). *Semiannual Update of the Status of New Reactor Licensing Activities and Future Planning for New Reactors.* SECY-06-0019, U.S. Nuclear Regulatory Commission, Washington, D.C.

[151] Richardson, S. D. & Reynolds, J. M. (2000). "An Overview of Glacial Hazards in the Himalayas." *Quaternary International, 65/66*, 31-47.

[152] Rong, Y., Jackson, D. D. & Kagan, Y. Y. (2003). "Seismic Gap and Earthquakes." *Journal of Geophysical Research, 108, 6-1-6-14.*

[153] Rothwell, R. G., Kenyon, N. H. & McGregor, B. A. (1991). "Sedimentary Features of the South Texas Continental Slope as Revealed by Side-Scan Sonar and High-Resolution Seismic Data." *American Association of Petroleum Geologists Bulletin, 75*, 298-312.

[154] Rudnicki, J. W. & Wu, M. (1995). "Mechanics of Dip-Slip Faulting in an Elastic Half-Space." *Journal of Geophysical Research, 100(B11)*, 22,173-22186.

[155] Saffer, D. M., Frye, K. M., Marone, C. & Mair, K. (2001). "Laboratory Results Indicating Complex and Potentially Unstable Frictiorial Behavior of Smectite Clay. *Geophysical Research Letters, 28(12)*, 2297-2300.

[156] Satake K. & Kanamori, H. (1991). "Abnormal Tsunamis Caused by the June 13, 1984, Torishima, Japan, Earthquake." *Journal of Geophysical Research, 96*, 19,933-19,939.

[157] Satake K. & Kato, Y. (2001). "The 1741 OshimaOshima Eruption: Extent and Volume of Submarine Debris Avalanche." *Geophysical Research Letters, 28(3)*, 427-430.

[158] Satake K. & Tanioka, Y. (2003). "The July 1998 Papua New Guinea Earthquake: Mechanism and Quantification of Unusual Tsunami Generation." *Pure and Applied Geophysics, 160*, 2087-2118.

[159] Satake K., Shimazaki, K., Tsuji, Y. & Ueda, K. (1996). "Time and Size of a Giant Earthquake in Cascadia Inferred from Japanese Tsunami Record of January 1700." *Nature, 379(6562)*, 246-249.

[160] Satake, K., Smith, J. R. & Shinozaki, K. (2002). "Volume Estimate and Tsunami Modeling for the Nuuanu and Wailau Landslides." In: *Hawaiian Volcanoes: Deep Underwater Perspectives,* E. Takahashi, P.W. Lipman, M.O. Garcia, J. Naka, and S. Aramaki (eds.), Geophysical Monograph 128, American Geophysical Union, pp. 333-348.

[161] Satake, K., Wang, K. & Atwater, B. F. (2003). "Fault Slip and Seismic Moment of the 1700 Cascadia Earthquake Inferred from Japanese Tsunami Descriptions." *Journal of Geophysical Research, 108(B11)*, 2535. doi: 10.1019/2003JB002521.

[162] Scholz, C. H. (1990). *The Mechanics of Earthquakes and Faulting.* Cambridge University Press, Cambridge, Massachusetts.

[163] Schreiner, L. C. & Riedel, J. T. (1978). *Probable Maximum Precipitation Estimates, United States East of the 105th Meridian.* NOAA

Hydrometeorological Report 51, National Weather Service, Washington, D.C.

[164] Schwab, W. C., Danforth, W. W., Scanlon, K. M. & Masson, D. G. (1991). "A Giant Slope Failure on the Northern Insular Slope of Puerto Rico." *Marine Geology, 96*, 237-246.

[165] Schwab, W. C., Danforth, W. W. & Scanlon, K. M. (1993). "Tectonic and Stratigraphic Control on a Giant Submarine Slope Failure: Puerto Rico Insular Slope." In: *Submarine Landslides: Selected studies in the U.S. Exclusive Economic Zone,* W.C. Schwab, H.J. Lee, and D.C. Twichell (eds.). U.S. Geological Survey Bulletin 2002, 60-68.

[166] Seed, H. B., Lee, K. L. & Makdisi, F. I. (1975). "The Slides in the San Fernando Dams During the Earthquake of February 9, 1971." American Society of Civil Engineers (ASCE), *Journal of Geotechnical Engineering, 101(GT7), 651-689.*

[167] Shuto, N. (1991). "Numerical Simulation of Tsunamis." In: *Tsunami Hazard,* E. Bernard (ed.), Kluwer Academic Publishers, Dordrecht, The Netherlands, pp. 171-191.

[168] Somerville P., Irikura, K., Graves, R., Sawada, S., Wald, D., Abrahamson, N., Iwasaki, Y., Kagawa, T., Smith, N. & Kowada, A. (1999). "Characterizing Crustal Earthquake Slip Models for the Prediction of Strong Ground Motion." *Seismological Research Letters, 70*, 59-80.

[169] Soofi, M. A. & King, S. D. (2002). "Oblique Convergence Between India and Eurasia." *Journal of Geophysical Research, 107(B5)*, Art. No. 2101. Doi: 10.1029/2001JB000636.

[170] Stein, S. & Okal, E. A. (2007). "Ultralong Period Seismic Study of the December 2004 Indian Ocean Earthquake and Implications for Regional Tectonics and the Subduction Process." *Bulletin of the Seismological Society of America, 97(1A)*, S279-5295.

[171] Sykes, L. R. (1971). "Aftershock Zones of Great Earthquakes, Seismicity Gaps, and Earthquake Prediction for Alaska and the Aleutians." *Journal of Geophysical Research, 76*, 8021-8041.

[172] Tanioka, Y. & Satake, K. (1996). "Tsunami Generation by Horizontal Displacement of Ocean Bottom." *Geophysical Research Letters, 23(8)*, 861-864.

[173] Tanioka, Y. & Seno, T. (2001). "Detailed Analysis of Tsunami Waveforms Generated by the 1946 Aleutian Tsunami Earthquake." *Natural Hazards and Earth System Sciences, 1,171-175.*

[174] ten Brink, U. S., Geist, E. L. & Andrews, B. D. (2006). "Size Distribution of Submarine Landslides and its Implication on Tsunami

Hazard in Puerto Rico." *Geophysical Research Letters, 33*, Art. No. L11307. doi: 10.1029/2006GL026125.

[175] ten Brink, U., Twichell, D., Geist, E., Chaytor, J., Locat, J., Lee, H., Buczkowski, B., Barkan, R., Solow, A., Andrews, B., Parsons, T., Lynett, P., Lin, J. & Sansoucy, M. (2008). *Evaluation of Tsunami Sources with the Potential to Impact the U.S. Atlantic and Gulf Coasts.* An Updated Report to the Nuclear Regulatory Commission by Atlantic and Gulf of Mexico Tsunami Hazard Assessment Group. U.S. Geological Survey Administrative Report, August 2008.

[176] Tinti, S., Maramai, A. & Cerutti, A.V. (1999). "The Miage Glacier in the Valley of Aosta (Western Alps, Italy) and the Extraordinary Detachment Which Occurred on August 9, 1996." *Physics and Chemistry of the Earth, Part A: Solid Earth and Geodesy, 24*, 157-161.

[177] Tinti, S., Bortolucci, E. & Romagnoli, C. (2000). "Computer Simulations of Tsunamis due to Sector Collapse at Stromboli, Italy. " *Journal of Volcanology and Geothermal Research, 96*,103128.

[178] Titov, V. V. (1997). *Numerical Modeling of Long Wave Runup.* Ph.D. Thesis, University of Southern California, Los Angeles, California.

[179] Titov, V. V. & Gonzalez, F. I. (1997). *Implementation and Testing of the Method of Splitting Tsunami (MOST) Model.* NOAA Technical Memorandum ERL PMEL-112, Pacific Marine Environmental Laboratory, Seattle, Washington.

[180] Titov, V. V. & Synolakis, C. E. (1997). "Extreme Inundation Flows During the Hokkaido-NanseiOki Tsunami." *Geophysical Research Letters, 24(11)*, 1315-1318.

[181] Titov, V. V. & Gonzalez, F. I. (2001). "Numerical Study of the Source of the July 17, 1998 PNG Tsunami." In *Tsunami Research at the End of a Critical Decade,* G.T. Hebenstreit (ed.), Kluwer Academic Publishers, Dordrecht, The Netherlands, pp. 197-207.

[182] Titov, V. V., Mofjeld, H. O., Gonzalez, F. I. & Newman, J. C. (1999). *Offshore Forecasting of Hawaiian Tsunamis Generated in Alaskan-Aleutian Subduction Zone.* NOAA Technical Memorandum ERL PMEL-114, Pacific Marine Environmental Laboratory, Seattle, Washington.

[183] Titov, V. V., Gonzalez, F. I., Bernard, E. N., Eble, M. C., Mofjeld, H. O., Newman, J. C. & Venturato, A. J. (2005). "Real-Time Tsunami Forecasting: Challenges and Solutions." In: *Developing Tsunami-Resilient Communities,* E.

[184] Bernard (ed.). The National Tsunami Hazard Mitigation Program, *Natural Hazards, 35(1)*, 4058.

[185] Tonkin, S., Yeh, H., Kato, F. & Sato, S. (2003). "Tsunami Scour Around a Cylinder." *Journal of Fluid Mechanics, 496*, 165-192.

[186] Trifunac, M. D., Hayir, A. & Todorovska, M. I. (2002). A Note on the Effects of Nonuniform Spreading Velocity of Submarine Slumps and Slides on the Near-Field Tsunami Amplitudes." *Soil Dynamics and Earthquake Engineering, 22*, 167-180.

[187] Trifunac, M. D., Hayir, A. & Todorovska, M. I. (2003). "A Note on Tsunami Caused by Submarine Slides and Slumps Spreading in One Dimension with Nonuniform Displacement Amplitudes." *Soil Dynamics and Earthquake Engineering, 23*, 223234.

[188] Tripsanas, E. K., Bryant, W. R., Prior, D. B. & Phaneuf, B. A. (2003). "Interplay Between Salt Activities and Slope Instabilities, Bryant Canyon Area, Northwest Gulf of Mexico." In: *Submarine Mass Movements and Their Consequences,* J. Locat and J. Mienert (eds.), Kluwer Academic Publishers, Dordrecht, the Netherlands, pp. 307315.

[189] Tsuji Y., K. Ueda, and K. Satake. 1998. "Japanese Tsunami Records from the January 1700 Earthquake in the Cascadia Subduction Zone." *Journal of the Seismological Society of Japan, 51, 1-17* [in Japanese with English title, abstract, and captions].

[190] Twichell, D. C., Valentine, P. C. & Parson, L. M. (1993). "Slope Failure of Carbonate Sediment on the West Florida Slope." In: *Submarine Landslides: Selected Studies in the U.S. Exclusive Economic Zone,* W.C. Schwab, H.J. Lee, and D.C. Twichell (eds.), U.S. Geological Survey Bulletin 2002, pp. 69-78.

[191] Ui, T., Takarada, S. & Yoshimoto, M. (2000). "Debris Avalanches." In: *Encyclopedia of Volcanoes,* H. Sigurdsson (ed.), Academic Press, San Diego, California, pp. 617-626.

[192] United Nations Educational, Scientific and Cultural Organization (UNESCO). (1997). *Numerical Method of Tsunami Simulation with the Leap-Frog Scheme.* International Union of Geodesy and Geophysics /International Oceanographic Commission (IUGG/IOC) TIME Panel, IOC Manuals and Guides No. 35, IOC.

[193] U.S. Geological Survey (USGS). (2007). *Geology of Interactions of Volcanoes, Snow, and Water: Mount St. Helens, Washington, Tsunami on Spirit Lake Early During 18 May 1980 Eruption.* URL: http://valcan.wr.usgs.gov/Projects/H2O+Volcarioe siFrozen/Geologv/MSH/MSH.tsunami.html, Accessed March 12, 2008.

[194] U.S. Nuclear Regulatory Commission (NRC). (1977). *Design Basis Floods for Nuclear Power Plants.* Regulatory Guide 1.59. Washington,

D.C.

[195] U.S. Nuclear Regulatory Commission (NRC). (1996). *Standard Review Plan for the Review of Safety Analysis Reports for Nuclear Power Plants.* LWR Edition, Draft Revision 3, Office of Nuclear Reactor Regulation, Washington, D.C.

[196] U.S. Nuclear Regulatory Commission (NRC). (2007). *Standard Review Plan for the Review of Safety Analysis Reports for Nuclear Power Plants.* LWR Edition, Office of Nuclear Reactor Regulation, Washington, D.C.

[197] Varnes, D. J. (1978). "Slope Movement Types and Processes." In: *Landslides: Analysis and Control,* R.L. Schuster and R.J. Krizek (eds.). National Academy of Sciences, Washington, D.C., pp. 1133.

[198] Vreugdenhil, C. B. (1994). *Numerical Methods for Shallow- Water Flow,* Kluwer Academic Publishers, Dordrecht.

[199] Waitt, R. B. & Pierson, T. C. (1994). "The 1980 (Mostly) and Earlier Explosive Eruptions of Mount St. Helens Volcano." In: *Geologic field trips in the Pacific Northwest,* D.A. Swanson and R.A. Haugerud, eds., Department of Geological Sciences, University of Washington, Volume 2, Chapter 21.

[200] Walder J. S., Watts, J., O.E. Sorensen, P. & Janssen, K. (2003). "Tsunamis Generated by Subaerial Mass Flows." *Journal of Geophysical Research, 108(B5)*, 2236. doi:10.1029/2001JB000707.

[201] Ward S. N. & Day, S. (2001). "Cumbre Vieja Volcano -- Potential Collapse and Tsunami at La Palma, Canary Islands." *Geophysical Research Letters, 28(17)*, 3397-3400.

[202] Ward, S. N. & Day, S. (2003). "Ritter Island Volcano Lateral Collapse and the Tsunami of 1888." *Geophysical Journal International, 154*, 891-902.

[203] Watts, P. (1998). "Wavemaker Curves for Tsunamis Generated by Underwater Landslides." *Journal of Waterway, Port, Coastal, and Ocean Engineering, 124(3)*, 127-137.

[204] Watts P. & Waythomas, C. F. (2003). "Theoretical Analysis of Tsunami Generation by Pyroclastic Flows." *Journal of Geophysical Research, 108(B12)*, 2563. doi:10.1029/2002JB002265.

[205] Waythomas, C. F. & Neal, C. A. (1998). "Tsunami Generation by Pyroclastic Flow During the 3500- year B.P. Caldera-Forming Eruption of Aniakchak Volcano, Alaska." *Bulletin of Volcanology, 60*, 110-124.

[206] Wesnousky, S. G. (1994). "The Gutenberg-Richter or Characteristic Earthquake Distribution; Which Is It?" *Bulletin of the Seismological Society of America, 84*, 1940-1959.

[207] Wilson, C. K., Long, D. & Bulat, J. (2003). "The Afen Slide—A Multi-Stage Slope Failure in the Faroe-Shetland Channel " In: *Submarine Mass Movements and Their Consequences,* J. Locat and J. Mienert (eds.), Kluwer Academic Publishers, Dordrecht, the Netherlands, pp. 317-324.

[208] Wynn, R. B. & Masson, D. G. (2003). "Canary Islands Landslides and Tsunami Generation: Can We Use Turbidite Deposits to Interpret Landslide Processes?" In: *Submarine Mass Movements and Their Consequences,* J. Locat and J. Mienert (eds.), Kluwer Academic Publishers, Dordrecht, the Netherlands, pp. 325-332.

[209] Wyss, M. (1979). "Estimating Maximum Expectable Magnitude of Earthquake from fault Dimensions." *Geology, 7, 336-340.*

[210] Yeh, H., Robertson, I. & Preuss, J. (2005). *Development of Design Guidelines for Structures that Serve as Tsunami Vertical Evacuation Sites.*

[211] Washington Division of Geology and Earth Resources, Open File Report 2005-4. Yomogida K. (1988). "Crack-Like Rupture Process Observed in Near-Fault Strong Motion Data." *Geophysical Research Letters, 15, 1223-1226.*

[212] Yoshioka, S., Hashimoto, M. & Hirahara, K. (1989). "Displacement Fields due to the 1946 Nankaido Earthquake in a Laterally Inhomogeneous Structure with the Subducting Philippine Sea plate — A Three-Dimensional Finite Element Approach." *Tectonophysics, 159*, 121-136.

[213] Zhao, S., Muller, R. D., Takahashi, Y. & Kaneda, Y. (2004). "3-D Finite-Element Modeling of Deformation and Stress Associated with Faulting: Effects of Inhomogeneous Crustal Structures." *Geophysical Journal International, 157*, 629-644.

Appendix A. Tsunami Hazard Assessment at a Hypothetical Nuclear Power Plant Site

This appendix provides a set of guidelines that will be applicable to most nuclear power plant sites in the United States. It presents the primary steps required to carry out tsunami hazard assessment at a hypothetical nuclear power plant site. These steps are not all-encompassing because many of the activities and analyses that need to be carried out are site-dependent and site-specific.

A.1. Preliminary Data Collection and Site Screening

The tsunami hazard assessment at a nuclear power plant is carried out using a hierarchical approach (see Chapter 1 of this report). The first step of this approach is to establish whether the region is subject to tsunamis. If it is found that the general region is subject to tsunamis, a further search for the site itself is conducted to determine the effect of the tsunamis on the site itself.

A.1.1. Is the Site Region Subject to Tsunamis?

A search of the National Geophysical Data Center (NGDC) tsunami database is carried out to determine whether any historical tsunami events were reported at or near the site. The NGDC database can be searched using the following parameters

- Runup Country: USA
- Runup State: the state where the proposed nuclear power plant is to be located

at http://www.ngdc.noaa.gov/nndc/struts/form?t=101650&s=167&d=166. The NGDC database also can be searched for tsunami sources that may affect the site using the NGDC search at *http://www.ngde.noaa.gov/ nndc/struts/form?t=101650&s=70&d=7* using similar search parameters.

At the time this report was written, a search of the NGDC database for runups in California returned 503 records. A similar search for Florida, New Jersey, and Texas returned 9, 20, and 3 records, respectively. The NGDC search of the tsunami sources for California returned 130 records. The corresponding tsunami record search for Florida, New Jersey, and Texas returned 5, 11, and 3 records. The runup locations are plotted in Figure A-1. Erroneous and doubtful tsunami runup reports are included in the NGDC database and may show up in the searches. For example, of the 503 tsunami runup records returned for California, 24 are labeled as being erroneous, 28 are labeled as being very questionable, and 14 are labeled as being questionable. Similarly, of the 130 records for tsunami runups in California, 18 are labeled as being erroneous, 20 are labeled as being very questionable, and 14 are labeled as being questionable. The tsunami source locations are plotted in Figure A-2.

The NGDC tsunami database also contains records and sources of inland tsunami-like waves that have occurred in lakes and rivers. A search for tsunami runups in Washington State returned 88 records, including two waves in Spirit Lake caused by the debris flow from the collapse of Mount St. Helen's dome during its 1980 eruption. A search for tsunami mnups in Missouri returned two events, both related to the 1811-1812 New Madrid earthquakes that caused waves in the Mississippi River.

The NGDC database also reports seiches induced by tsunamis and meteorological forcing. For example, a tsunami runup search for Michigan returned four events; the source for one, a ground swell or wave in the Detroit River, is listed as unknown, and the other three events are attributed to meteorologically induced seiches in Lake Huron or to an offshore landslide.

Based on results from searches of the NGDC tsunami database, it can be determined if the general region has been subjected to tsunamis or to tsunami-like waves.

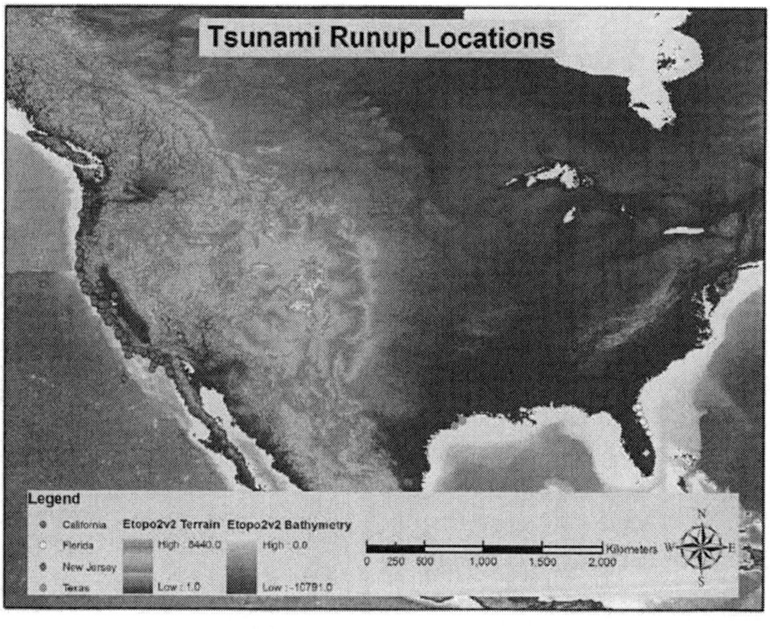

Figure A-1. Tsunami runup locations in selected states (from the NGDC tsunami database).

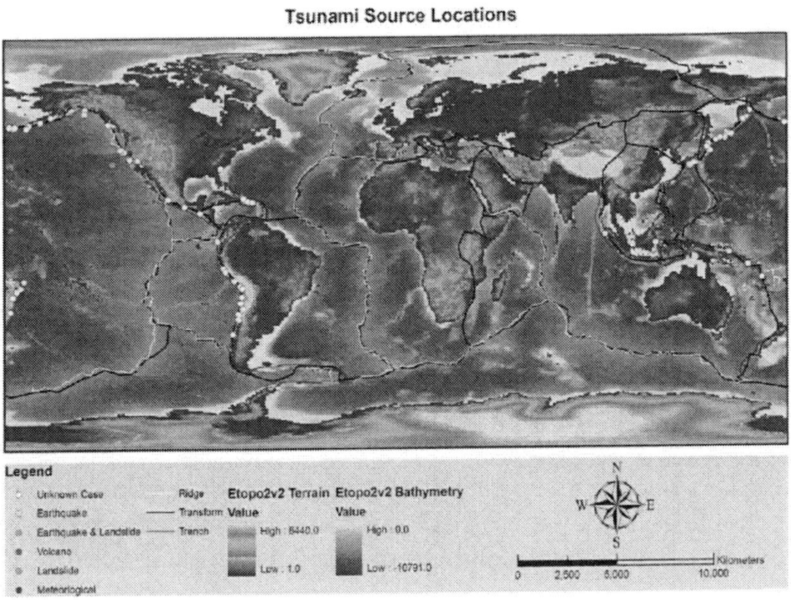

Figure A-2. Tsunami source locations for reported runups in selected states.

A.1.2. Is the Plant Site Affected by Tsunamis?

Based on data from the NGDC tsunami database, an assessment is made regarding the degree to which the proposed site itself is affected by tsunamis. It is possible that even though the region is subject to tsunamis, the site itself may be safe. This is essentially a site screening step.

Historical tsunami data and other geophysical, topographical, and hydrological data are necessary to carry out site-screening. These data include the following:

1. List of historical tsunamis and associated runups
2. Map of the proposed site and its vicinity
3. Map of potential tsunamigenic sources
4. Topographic and bathymetric maps
5. Geologic and geotechnical maps
6. Hydrological setting of the site
7. Onshore and offshore soil and substrate characteristics.

The list of historical tsunamis, the associated wave characteristics, and information regarding their sources should already be available from the search conducted to identify tsunamis in the general region of the proposed nuclear power plant. The location of the proposed nuclear power plant site and its relation with the physiographic characteristics of the region should be evaluated from a map that has sufficient resolution. The topographic and bathymetric maps provide baseline data for interpretation of runup characteristics of the tsunami waves. The hydrological setting of the site allows the interpretation of the interaction of tsunami waves with nearby water bodies such as an estuary, a natural or man-made embayment, or other coastal structures.

The above data should be critically evaluated to answer the following questions:

1. Can a tsunami wave run up to the proposed grade elevation of the nuclear power plant?
2. Can tsunami waves inundate a sufficient distance from the shoreline to affect proposed facilities of the nuclear power plant?
3. Can tsunami waves interact with near-shore geometry to result in resonance?
4. Can tsunami waves be expected to lead to severe erosion near the proposed nuclear power plant facilities?
5. Can tsunami waves severely affect local communities and infrastructure?

It may be necessary to assemble an interdisciplinary team of experts to evaluate the data and to answer the above questions. At this site-screening stage, it may not be necessary to precisely estimate the impacts from tsunamis quantitatively. However, a thorough qualitative description can result in a very good estimate of the amount of effort required to carry out a complete tsunami hazard analysis. This evaluation can also indicate if the site is exposed to severe tsunami, hazards and if the considerable effort to carry out a subsequent complete tsunami hazard assessment is worthwhile.

Based on the evaluation of the degree of impact from tsunamis at the proposed site, a decision can be made to proceed with a detailed tsunami hazard assessment to provide the bases for the design and the protection of the SSC of the plant.

A.2. Site Investigation and Data Collection

To proceed with a detailed tsunami hazard assessment at the proposed nuclear power plant site, the preliminary data collected for site screening needs to be supplemented by additional site investigation and data collection. These data collection efforts may include paleo-tsunami investigations, compilation of bathymetry and topography, collection of high-resolution nearshore bathymetry, characterization of geophysical properties of the nearshore substrate, and tide or other water level records.

A.2.1. The Tsunami Record

If the historical tsunami record at the proposed site is relatively short, a paleo-tsunami investigation to determine the existence and severity of prehistorical tsunamis should be considered. As a first step, paleo-tsunami databases such as the one developed by Peters et al. (2003) should be searched to locate data that may already be available from previous efforts. Paleo-tsunami experts should be contacted to determine whether more recent data collection efforts are underway.

If paleotsunami data are not readily available for the proposed nuclear power plant site, a site-specific paleo-tsunami investigation should be considered. Chapter 5 of this report lists the pieces of information measured or interpreted from the paleo-tsunami deposits. It may be necessary to obtain expert help to carry out paleotsunami surveys, data collection, and interpretation.

A.2.2. Bathymetry and Topography

For simulation of the propagation of the PMT waves and the inundation caused by the waves, good- quality bathymetry and topography data are needed. A search should be conducted to determine whether a combined bathymetry and topography data set is available for the proposed nuclear power plant site or its vicinity. This search may save much of the effort needed to create this data set from raw bathymetry and topography data sources (see Chapter 5 of this report).

The extent of spatial coverage needed for tsunami simulation depends on several factors: (1) whether the tsunami source is near-field or far-field, (2) whether pre-computed tsunami wave heights from far-field sources are available (e.g., from the National Oceanic and Atmospheric Administration Center for Tsunami Research [NOAA NCTR] database), and (3) the geometry of the shoreline. Global bathymetry datasets may be used for tsunami propagation modeling from far-field sources if precomputed tsunami wave heights from these sources are not available. The use of a global bathymetry data set may also be necessary in near-shore inundation modeling that uses the NOAA NCTR pre-computed offshore tsunami wave heights depending on the offshore distance at which these wave heights are provided.

The resolution of the near-shore bathymetry and topography data may be critical for accurate simulation of the propagation and inundation of the tsunami waves. Available data sources described in Chapter 5 of this report should be used to collect the best available bathymetry and topography data from the NGDC, the National Ocean Service (NOS), and the U.S. Geological Survey (USGS). Airborne remote-sensing technologies, like LiDAR and IfSAR, provide the best quality topographic data. Topographic and bathymetric data derived from Light Detection And Ranging (LiDAR) and Interferometric Synthetic Aperture Radar (IfSAR) systems can be obtained from the NOAA Coastal Services Center (CSC) (see Chapter 5 of this report). In absence of NOAA CSC coverage, it may be necessary to carry out dedicated data collection and processing.

A high-quality combined bathymetry and topography data set should be created by merging the available onshore and offshore elevation information. Several spatial interpolation techniques are available in geographic information system (GIS) software and can be used to create grids from irregularly spaced point data or from elevation contour lines. The final grid for tsunami modeling should be created with all cells set to non-zero elevation values because an elevation value of zero could cause anomalies in some tsunami inundation models (Gonzalez et al. 2006). A solution suggested by Gonzalez et al. (2006) is to use a small positive value for land grid cells (e.g., 0.01 m) and a correspondingly small negative value for water grid cells (e.g., -0.01 m). The final Digital Elevation Model (DEM) grid should be checked for consistency using known features on land and offshore.

A.2.3. Geophysical Data

Bottom pressure gauge data from available locations within the tsunami propagation and inundation regions should be obtained from the NGDC. Water column height data obtained for the month of January 2007 at a DART station off the coast of San Francisco are shown in Figure A-3.

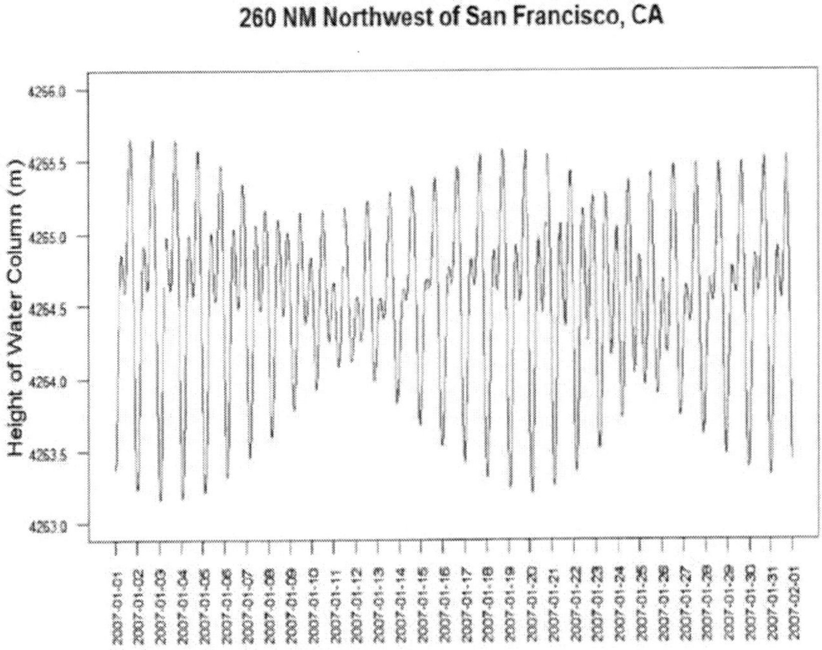

Figure A-3. DART water column height data for January 2007 at Station 46411.

Tides or water surface elevation data are needed to specify antecedent conditions before the arrival of the tsunami. Historical tide data available from the NOAA Center for Operational Oceanographic Products and Services (CO-OPS) Tides and Currents website should be obtained. For example, Figure A-4 shows the tide data for the month of January 2007 at the San Francisco station.

A statistical analysis of historical tide data should be carried out to determine the appropriate tide levels at specified exceedance probabilities. For example, the low tide level at the 90-percent exceedance probability level could be used for the antecedent water level to determine the minimum tsunami water surface elevation during rundown, and the high tide level at the 10-percent exceedance probability level could be used as the antecedent water level to determine the maximum tsunami water surface elevation during runup.

Tide-level data are also useful during validation of tsunami models. The data can be used to verify whether the model predictions of tsunami wave heights compare well with observed water levels at these locations. If possible, data from several tide gauges at different locations along the propagation path of a few historical tsunamis should be used to assess the accuracy of the simulations.

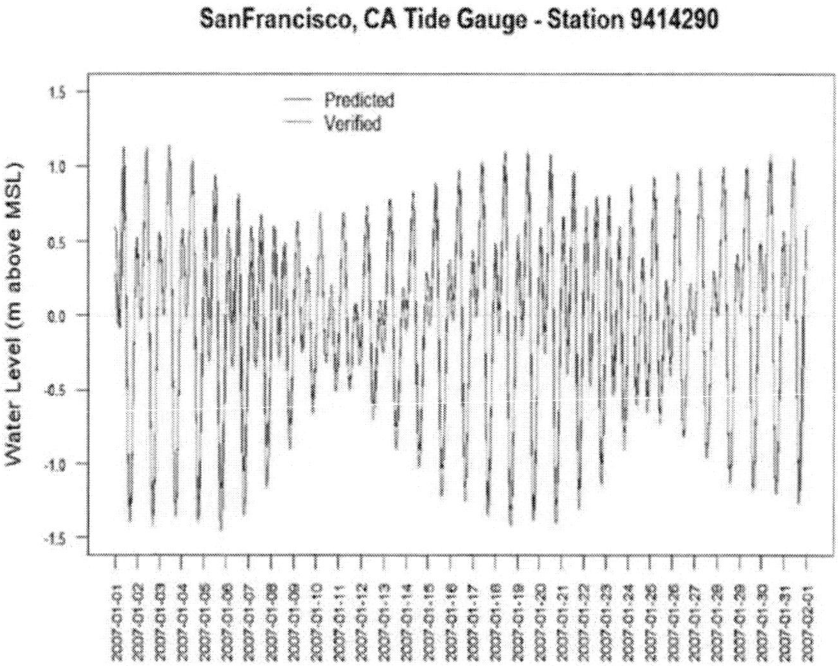

Figure A-4. Tide-level data at San Francisco station 9414290.

Table A-1. Search Results from the NGDC Grain Size Database

Dataset	Contributor	Scientist	Ship	Cruise
01995001	LDEO†	Horn	Robert Conrad	RC10
09005007	NAVOCEANO‡		Rehoboth	D-5
09005039	NAVOCEANO	Loomis	Unknown	L.I. 446
09005043	NAVOCEANO	Loomis	Unknown	L.I. 450

†Lamont-Doherty Earth Observatory

‡U.S. Naval Civil Engineering Laboratory

Information regarding soil grain size distribution should be obtained from the NGDC. A search of the NGDC grain-size database for the bounding rectangle ranging in latitudes from 32°N to 44°N and ranging in longitudes from 122°W to 130°W returned four cruises and 22 samples. Table A-1 lists the results of this search.

Three tab-delimited files can be exported from the database search that contain cruise and sample data, interval data, and weighted percentage data for each data set. The first few records of the cruise and sample export file from a data set are shown in Table A-2.

The locations of these samples should be plotted on a map to determine whether they are appropriate for use in tsunami hazard analysis at the proposed nuclear power plant site. If the number of samples is found to be insufficient or the locations of the samples are found to be inappropriate to derive any conclusions about the grain-size distributions near the proposed nuclear power plant site, a site-specific data collection effort may be needed to determine these properties.

Table A-2. Exported Cruise and Sample File from the NGDC Grain-Size Database Search for Dataset ID 01995001

```
#*****************************************************
#USDOC/NOAA/NESDIS/National Geophysical Data Center (NGDC)
#World Data Center for Marine Geology and Geophysics, Boulder
#Seafloor Sediment Grain Size Database
#Data Set WDCMGG00127
#URL http://www.ngdc.noaa.gov/mgg/geology/size.html
#file created: 12-MAR-07 as a sample level export
#questions to: Carla.J.Moore@noaa.gov
#phone:     303-497-6339 fax: 303-497-6513
#address:   NOAA/National Geophysical Data Center
#           325 Broadway Code E/GC3
#           Boulder, Colorado USA 80305-3328
#*****************************************************
#-------------------------------------------------------
#           MGG01995001
#-------------------------------------------------------
# Institution: Lamont-Doherty Earth Observatory (LDEO)
#      Title: Acoustic Provinces of the North Pacific Based on Deep Sea Cores
#     Report: Technical Report No. 3, CU-3-67 NAVSHIPS N00024-67-C-1186
#     Set id: FE00502
#    Authors: D.R. Horn
#      Ships: Vema, Robert Conrad
#    Cruises: 20,21,10
#      Areas: North Pacific
#    Funding: Navy-misc.
#    Project:
# # of samples: 248
# Cruise dates: April 1,1964- August 31,1966
#    Devices: 248 cores
#   Analyses: 248 texture, 245 engineering properties.
#    Formats: Available in .pdf form. Grain size data in digital form
# # microfiche: 2
# # pages paper: 49
#-------------------------------------------------------

mggid    ship          cruise sample  device        date_collected lat     lon      water_depth core_length sample_comments
01995001 Robert Conrad  RC10  RC10156 core, piston 19660516  22.3333 157.8167 5402
01995001 Robert Conrad RC10   RC10157 core, piston 19660517  24.7667 159.1333 5682
01995001 Robert Conrad RC10   RC10158 core, piston 19660518  28.1167 160.6    5892
01995001 Robert Conrad RC10   RC10159 core, piston 19660519  31.2167 162.3    5894
01995001 Robert Conrad RC10   RC10160 core, piston 19660520  32.4667 159.8333 4621
01995001 Robert Conrad RC10   RC10161 core, piston 19660521  33.0833 158      3587
01995001 Robert Conrad RC10   RC10162 core, piston 19660522  31.4167 158.8    3913
01995001 Robert Conrad RC10   RC10163 core, piston 19660522  32.7167 157.5    3550
01995001 Robert Conrad RC10   RC10164 core, piston 19660523  31.7333 157.5    3766
01995001 Robert Conrad RC10   RC10166 core, piston 19660523  31.8333 157.3333 3792
....
....
```

Soil properties like buoyant specific weight, coefficient of consolidation, and others may be required to determine the hazard from tsunamis. Because such data usually is not expected to be available from repositories, it would need to be obtained using site-specific surveys. The properties that need to be determined from site-specific investigation will depend on the hazard indices being estimated or hazard models being used.

The NGDC also archives data obtained during the Ocean Drilling Program and the Deep Sea Drilling Project. These data sets contain information about grain-size distribution, particle density, porosity, and other geophysical and geochemical parameters. These data sets should also be investigated to determine their applicability to the proposed site.

A.2.4. Tsunami Sources

Three distinct categories of tsunami sources should be considered for the proposed nuclear power plant site:

1. Known sources of historical tsunamis
2. Potential sources determined by a paleotsunami study
3. Sources that may generate a tsunami-like wave in a nearby inland water body.

A search of the NGDC tsunami source database should be conducted to determine the sources of historical tsunami events at and near the proposed nuclear power plant site. As an example, Figure A-2 shows the tsunami source locations for the runup events for the states of California, Texas, Florida, and New Jersey. The data from the NGDC database should be processed to exclude erroneous entries, and tsunami experts should be consulted to determine whether any of the questionable tsunami sources are credible for the proposed nuclear power plant site.

If a paleotsunami study was carried out, either a review of an existing effort or a site-specific one, conclusions regarding potential tsunami sources that may affect the proposed site and its vicinity should be considered.

The hydrological and topographical setting of the site should be thoroughly evaluated to determine whether a tsunami-like wave in a nearby water body may result in a hazard at the proposed site. The mechanisms that may generate a tsunami-like wave in nearby inland water bodies include hillslope failures (similar to subaerial landslide), subaqueous landslides, and earthquakes. These sources should be comprehensively documented with regard to their tsunamigenic potential.

A.3. Specification of the PMT

The PMT is the most severe tsunami that can reasonably occur at a site. Because the severity of tsunamis that reach a site of interest is controlled not only by the severity of the source mechanism but also by the behavior of the tsunami waves during propagation and inundation, it is generally not possible to a priori determine the PMT source (González et al. 2007) for a specified site. However, parameters of similar sources can be used to determine several candidate PMTs. These candidate PMTs should be further evaluated to determine the most severe impact at the site.

Three source mechanisms of tsunami generation need to be considered: earthquakes, landslides (subaerial and submarine), and volcanoes. Table A-3 lists the source parameters for the three source mechanisms that can be used to determine the most severe source for the respective mechanisms.

The most severe tsunamigenic earthquake should be determined by evaluating both near and far-field earthquake zones. For example, for the West Coast of the United States, earthquake zones including the Cascadia subduction zone (where the Juan de Fuca Plate is subducting below the North American Plate; see Figure A-5), and subduction zones marked by the Aleutian trench (where the Pacific Plate is subducting below the North American Plate), the Kurile trench (where the Pacific Plate is subducting below the Okhotsk Plate), the Japan trench (where the Pacific Plate is subducting below the Eurasian plate), the Philippine trench (a complex tectonic activity zone where the Eurasian and the Philippine plates are converging), the Bougainville trench (northeast of Bougainville, the largest of the Solomon Islands and a province of Papua New Guinea, where several micro-plates are interacting with the Pacific Plates), and the Peru-Chile trench (where the Nazca Plate is subducting below the South American Plate) should be investigated. The maximum moment magnitude that can reasonably be expected to occur at the relevant earthquake zones should be determined. For each earthquake zone, a list of candidate PMT earthquakes should be compiled with conservatively estimated source parameters that maximize the tsunamigenic potential of the earthquakes. For example, an earthquake that results in a greater vertical displacement will have a greater tsunamigenic potential than one with a smaller vertical displacement, even if both had the same magnitude. Similarly, an earthquake closer to the site may result in a larger tsunami because dispersion of the generated waves could be minimal. The relative orientation of the rupture zone to the site can significantly affect the directivity of the tsunami waves; therefore, a smaller earthquake that is optimally oriented to the proposed site can cause a more severe tsunami at the site.

Landslides often cause tsunamis that are locally severe but may be significantly weaker in the far-field. Therefore, the proximity of a tsunamigenic landslide to the site is very important. The stability of a slope should be considered to determine possible landslide locations that may pose a tsunami hazard to the proposed site. The tsunamigenic parameters for a landslide source are the volume, the speed and acceleration of the slide, and the cohesiveness of the slide material. Conservative estimates of source parameters should be made. Historical or known landslides as well as potential landslides should be evaluated based on their respective tsunamigenic source characteristics. The landslides that have the greatest potential to cause a severe tsunami should be selected in the candidate PMT landslides.

Tsunamis generated by volcanic activity in the vicinity of the proposed site should be evaluated, and their tsunamigenic source parameters (listed in Table A-3) should be conservatively estimated. Tsunamis from far-field sources may be of concern to the proposed site if the activity involves possible caldera collapse or large flank failures. These far-field volcanoes should be evaluated with respect to their tsunamigenic potential and the possibility of the generated tsunamis to reach the site. Based on an evaluation of their tsunamigenic potential, candidate PMT volcanic sources should be selected.

All candidate PMT sources listed using the procedure described above should be used in tsunami generation, propagation, and inundation simulations. Hazards from all of these tsunami simulations should be evaluated to arrive at the hazards posed to the proposed site, as described below.

Table A-3. Source Parameters that Control the Severity of Generated Tsunamis

Source Mechanism	Source Parameters
Earthquakes	Magnitude, depth, vertical displacement, proximity to the site, orientation
Landslides	Proximity to the site, volume, slope stability, material properties, speed and acceleration
Volcanoes	Proximity to the site, VEI, pyroclastic flow properties, caldera size and depth

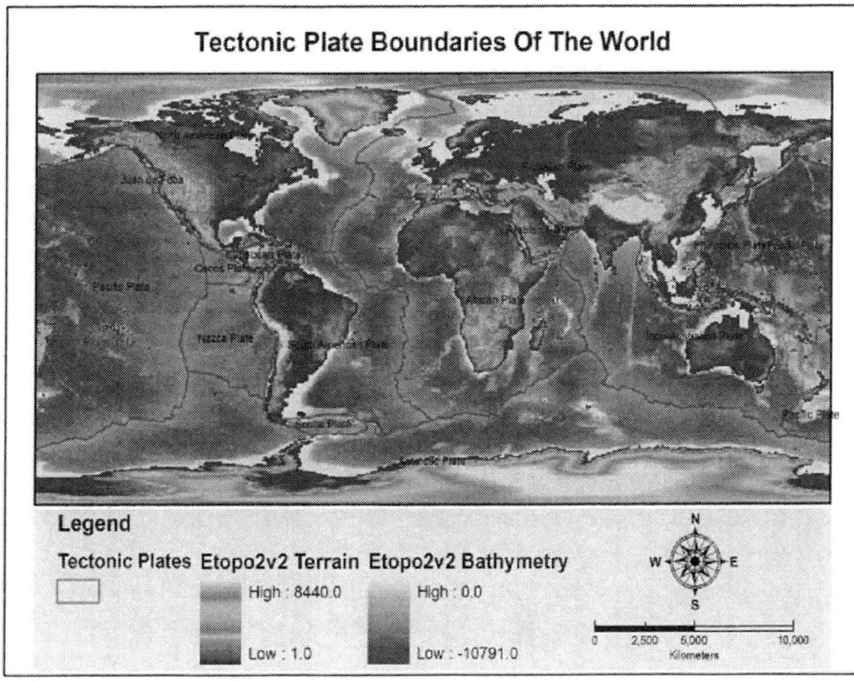

Figure A-5. Tectonic plate boundaries of the world.

A.3.1. Scenario Tsunamis

Scenario tsunamis are those obtained from a set of different source parameters for each postulated candidate PMT source. For example, for earthquake sources, the source parameters for each candidate PMT should be varied within a reasonable range that is realistically possible to obtain scenario tsunamis. The NCTR database, when available for the postulated earthquake-induced PMT, should be consulted to construct the scenario tsunamis by varying the location and number of unit sources. For each postulated earthquake-induced PMT, a set of tsunamis should be obtained using the NCTR tsunami-forecasting database approach. Each of these scenario tsunamis should be propagated to the site using site-specific tsunami-propagation models, and inundation effects should be modeled using site-specific inundation models. Tsunami hazards should be estimated for each scenario tsunami. The most severe hazard among the scenario tsunamis for all candidate PMTs for all tsunamigenic mechanisms should be specified as the site characteristic related to tsunamis.

A.4. Simulation of Tsunami Waves

For all scenario tsunamis, generation, propagation, and inundation modeling should be carried out. It may be possible to use the precomputed offshore tsunami wave heights from the NCTR tsunami forecasts database to set boundary conditions for inundation models for earthquake sources. However, if the NCTR forecast database does not contain precomputed scenarios applicable to the proposed site, a complete tsunami generation, propagation, and inundation modeling study may be required. Precomputed offshore tsunami wave heights for sources other than earthquakes may not be feasible. Tsunamis from these sources would need to be simulated on a case-by-case basis.

A tsunami modeling system, capable of modeling the three phases, generation, propagation, and inundation, should be used to simulate the scenario tsunamis from each candidate PMT source selected above. The tsunami modeling system should reflect the state-of-the-art in tsunami science. The tsunami modeling system may be composed of different models that may be appropriate for the three phases of the tsunami. The modeling system may also include different models that are appropriate for different source mechanisms. However, all of the separate components of the modeling system should be properly validated and verified using appropriate data, or shown to be accurate using site-specific data. Standards for evaluating tsunami models are described by González et al. (2007, Appendix C). For tsunami hazard assessment at proposed nuclear power plant sites, criteria for evaluating tsunami models that specifically deal with verifying model predictions against historical tsunami observations are especially important. To reduce the effort involved with several separate validation and verification studies, use of a community-accepted modeling system is recommended.

A.4.1. Generation

For each scenario of the postulated candidate PMT earthquakes, the initial tsunami waveform should be estimated conservatively. For earthquake sources, the Okada (1985) formulation could be used to specify the initial waveform. If the postulated rupture is highly heterogeneous, the Okada (1985) formulation may be used with the rupture subdivided into several cells within which the slip distribution could be assumed to be uniform.

Landslide and volcanic sources may need source-specific models of mass movement and tsunami generation. Experts should be consulted to properly define and constrain tsunami generation models of these sources with help from source surveys and model validation. For both of these sources, the coupling of the generated tsunami wave with the dynamics of the mass movement should be carefully investigated. If the coupling does not result in matching the speed of the mass movement with the speed of the tsunami waves, the reasons for this mismatch should be documented.

It is also possible that concurrent tsunamigenic sources need to be considered during a tsunamigenic event. For example, a tsunamigenic earthquake may generate a tsunami by vertical displacement of the water column and also may cause a subaqueous landslide that subsequently generates another tsunami. These concurrent events also should be evaluated as a potential PMT event, especially if it is possible that the separate effects of the concomitant events may combine to produce a more severe tsunami. It is also possible that the initial waveforms from the concomitant tsunamigenic sources may be located at significant distances from each other in space as well as in time. The potentially different locations and directivity of concomitant tsunamis may result in significantly altered wavefields with potential impact on the required extent of the modeling domain.

A.4.2. Propagation

The propagation of the generated tsunami waves from the source location toward the site should also be modeled. The spatial extent of the modeling domain should be sufficient to account for the effects of the bathymetry, the underwater mounts and ridges, and the presence of islands. The boundary conditions should be carefully specified (see Chapter 6 of this report). For cases where concomitant tsunamis are postulated (such as when a tsunamigenic earthquake triggers a landslide that also generates a tsunami), the model simulations should be set up to account for the separate tsunami sources.

The precomputed offshore tsunami wave heights from the NCTR tsunami forecast database may be used to reduce the effort of setting up and simulating an extensive area from the source to the proposed site. The precomputed offshore tsunami wave heights may be used to specify boundary conditions for a significantly smaller modeling domain. Currently, only earthquake sources can be modeled using the unit source approach, as demonstrated by the NCTR tsunami forecast database.

The propagation model should be chosen based on its acceptance in the tsunami modeling community. If a model other than a widely accepted and validated one is used, the reason for this selection should be documented, and the performance of the model should be clearly demonstrated using observed tsunami wave heights and travel times.

The waves from the candidate PMT sources should be simulated for a sufficiently long duration to ensure that all potentially large waves in the train of the tsunami waves have been simulated.

A.4.3. Inundation

During the inundation phase, the nonlinearity of the governing equations cannot be ignored. A model that uses nonlinear governing equations should be used. Near-shore geometry, bottom friction, and interaction of the tsunami waves with offshore and onshore structures can significantly affect the behavior of the tsunami waves in the inundation phase and therefore have a significant impact on the estimation of the metrics describing the hazards from the tsunami waves. Therefore, the model used to simulate the inundation phase should be validated and verified against site-specific observations. If site-specific observations are not available at the proposed site, model verification at similar sites may be used.

The simulations of the inundation phase of the scenario tsunamis for each candidate PMT should be carried out to obtain the maximum runup elevation, the spatial extent of runup, the minimum drawdown elevation, the spatial extent of drawdown, and the velocity and momentum fields.

A.5. Estimation of Hazards from the PMT

Based on the results from the inundation model for each scenario tsunami for all postulated candidate PMTs, the most severe hazard metrics should be determined, including the following:

1. high-water level: a map of the highest water surface elevation
2. low-water level: a map of the lowest water surface elevation
3. potential for scouring: a map of the scour-enhancement parameter described in Chapter 5
4. deposition: a map of areas susceptible to deposition from tsunami wave action
5. forces: a map of the velocity field that may be used to estimate forces described in Chapter 5
6. debris accumulation: a map of areas susceptible to debris accumulation

7. projectiles: a map of the impact forces expected from a postulated worst-case projectile .

The maps obtained above should be specified as site characteristics related to tsunamis at the selected site. These site characteristics should be used to specify design bases for all SSC that are important to safety and that may be exposed to tsunami hazards.

REFERENCES

[1] Gonzalez F. I, Geist, E., Synolakis, C., Arcas, D., Bellomo, D., Carlton, D., Horning, T., Jaffe, B., Johnson, J., Kanoglu, U., M4eld, H., Newman, J., Parsons, T., Peters, R., Peterson, C., Priest, G., Titov, V., Venturato, A., Weber, J., Wong, F. & Yalciner, A. (2006). *Seaside, Oregon Tsunami Pilot Study—Modernization of FEMA Flood Hazard Maps,* NOAA OAR Special Report, NOAA/OAR/PMEL, Seattle, Washington.

[2] Gonzalez F. I., Bernard, E., Dunbar, P., Geist, E., Jaffe, B., Kanoglu, U., Locat, J., Mofjeld, H., Moore, A., Synolakis, C. & Titov, V. (2007). *Scientific and Technical Issues in Tsunami Hazard Assessment of Nuclear Power Plant Sites.* NOAA Technical Memorandum OAR PMEL-136, Pacific Marine Environmental Laboratory, Seattle, Washington.

[3] Okada U. (1985). "Surface Deformation due to Shear and Tensile Faults in a Half-Space,." *Bulletin of the Seismological Society of America, 75(4),*1135-1154.

[4] Peters B., Jaffe, B., Gelfenbaum, G. & Peterson, C. (2003). *Cascadia tsunami Deposit Database.* U.S. Geological Survey Open-File Report 03-13, and electronic database and GIS coverage [URL: htth://geopubs.wr.usgs.gov/open-File/of03-13/*1*. Accessed March 12, 2008.

In: Tsunami Hazard Assessment at U.S. Nuclear… ISBN: 978-1-61470-865-0
Editors: Roger F. Goldman

Chapter 2

U.S. TSUNAMI PROGRAMS: A BRIEF OVERVIEW

Peter Folger

SUMMARY

A 9.0 magnitude earthquake struck off Japan's northeast coast near Honshu in the afternoon on Friday, March 11, 2011 (12:46 a.m. eastern time in the United States). The earthquake triggered a tsunami that has caused widespread devastation to parts of the coastal regions in Japan closest to the earthquake. The tsunami traveled across the Pacific Ocean, and the National Oceanic and Atmospheric Administration (NOAA) tsunami warning centers in Hawaii and Alaska issued tsunami warnings for coastal areas of Hawaii, Guam, the Commonwealth of the Northern Marianas, American Samoa, Alaska, and California. Although the tsunami caused widespread damage along the northeast coast of Japan, tsunami warnings issued from the tsunami warning centers gave the above U.S. Pacific territories, Hawaii, and the U.S. West Coast adequate warning to prepare for incoming waves.

NOAA's National Weather Service (NWS) manages the two tsunami warning centers that monitor, detect, and issue warnings for tsunamis generated in the Pacific Ocean. The NWS operates the Pacific Tsunami Warning Center (PTWC) at Ewa Beach, HI, and the West Coast/Alaska Tsunami Warning Center (WC/AKTWC) at Palmer, AK. The National Tsunami Hazards Mitigation Program (NTHMP) assists states in emergency planning and in developing maps of potential coastal inundation for a tsunami of a given intensity. The goal of NTHMP is to ensure adequate advance warning of tsunamis along all the U.S. coastal areas and appropriate community response to a tsunami event.

The tsunami warning centers monitor and evaluate data from seismic networks and determine if a tsunami is likely based on the location, magnitude, and depth of an earthquake. If the center determines that a tsunami is likely, it transmits a warning message to NOAA's weather forecasting offices and state emergency management centers, as well as to other recipients. The centers monitor coastal water-level data, typically with tide-level gages, and data from NOAA's network of Deep-ocean Assessment and Reporting of Tsunamis (DART) detection buoys to confirm that a tsunami has been generated, and if not, to cancel any warnings. Shortly after the 2004 tsunami in the Indian Ocean, Congress passed the Tsunami Warning and Education Act (P.L. 109-424), to enhance and modernize the existing Pacific Tsunami Warning System to increase coverage, reduce false alarms, and increase the accuracy of forecasts and warnings, among other purposes. As a result, the array was expanded to a total of 39 DART buoys in March 2008.

Funding for the NOAA tsunami program supports three main categories of activities: (1) *warning*, such as the activities of the tsunami warning centers and DART network; (2) *mitigation*, such as the activities of NTHMP; and (3) *research*, including activities conducted by the Pacific Marine Environmental Laboratory and the National Buoy Data Center. The Government Accountability Office (GAO) noted that total funding for all these activities ranged from $5 million to $10 million annually between FY1997 and FY2004, but increased after the 2004 Indian Ocean tsunami from approximately $27 million in FY2005 to $42 million in FY2009. Funding in FY2010 was $41 million.

Currently, 7 of the 39 DART buoys are not operational. Of the 7 buoys that are not working, 5 are deployed in the Pacific Ocean. If more DART buoys fail, and regional forecasting capabilities are impaired, then the NOAA Administrator must notify Congress within 30 days. According to NOAA, the current continuing resolution (P.L. 112-4) does not allow the NWS to allocate FY2011 funding to purchase ship time required to repair the 7 DART buoys that are not working.

JAPAN EARTHQUAKE AND TSUNAMI

A 9.0 magnitude massive earthquake struck off Japan's northeast coast near Honshu in the afternoon on Friday, March 11, 2011 (12:46 a.m. eastern time in the United States). The earthquake triggered a tsunami[1] that has caused widespread devastation to parts of the coastal regions in Japan closest to the earthquake. The tsunami traveled across the Pacific Ocean, and the National Oceanic and Atmospheric Administration (NOAA) tsunami warning centers in Hawaii and Alaska issued tsunami warnings for coastal areas of Hawaii, Guam, the Commonwealth of the Northern Marianas, American Samoa, Alaska, and California. The first tsunami waves reached Hawaii in the early morning of March 11,[2] and reached the west coast of the United States later in the morning (Pacific time). Although the tsunami caused widespread damage along the northeast coast of Japan, tsunami warnings issued from the tsunami warning centers gave the above U.S. Pacific territories, Hawaii, and the U.S. West Coast adequate warning to prepare for incoming waves.[3] In addition, the long distance traveled across the Pacific from the earthquake epicenter attenuated the energy associated with the tsunami thousands of miles from its source. In contrast, the city of Sendai, Japan, is just 80 miles west of the epicenter.[4]

TSUNAMI WARNING CENTERS

NOAA's National Weather Service (NWS) manages the two tsunami warning centers that monitor, detect, and issue warnings for tsunamis generated in the Pacific Ocean. The NWS operates the Pacific Tsunami Warning Center (PTWC) at Ewa Beach, HI, and the West Coast/Alaska Tsunami Warning Center (WC/AKTWC) at Palmer, AK. The PTWC monitors for tsunamis and issues warnings for the Hawaiian Islands, the U.S. Pacific territories, and other U.S. and international interests in the Pacific Basin. The center was established in 1949, after a strong earthquake and massive landslides off the coast of southwest Alaska caused a disastrous tsunami for the Hawaiian Islands only hours later. The WC/AKTWC was established in 1967, following a magnitude 9.2 earthquake that struck Anchorage, AK, in 1964 and caused major earthquake and localized tsunami damages.[5] The WC/AKTWC is responsible for issuing tsunami warnings to emergency management officials in Alaska, British Columbia (Canada), Washington

State, Oregon, and California. The WC/AKTWC also serves as the center for warning U.S. populations located in the western Atlantic.

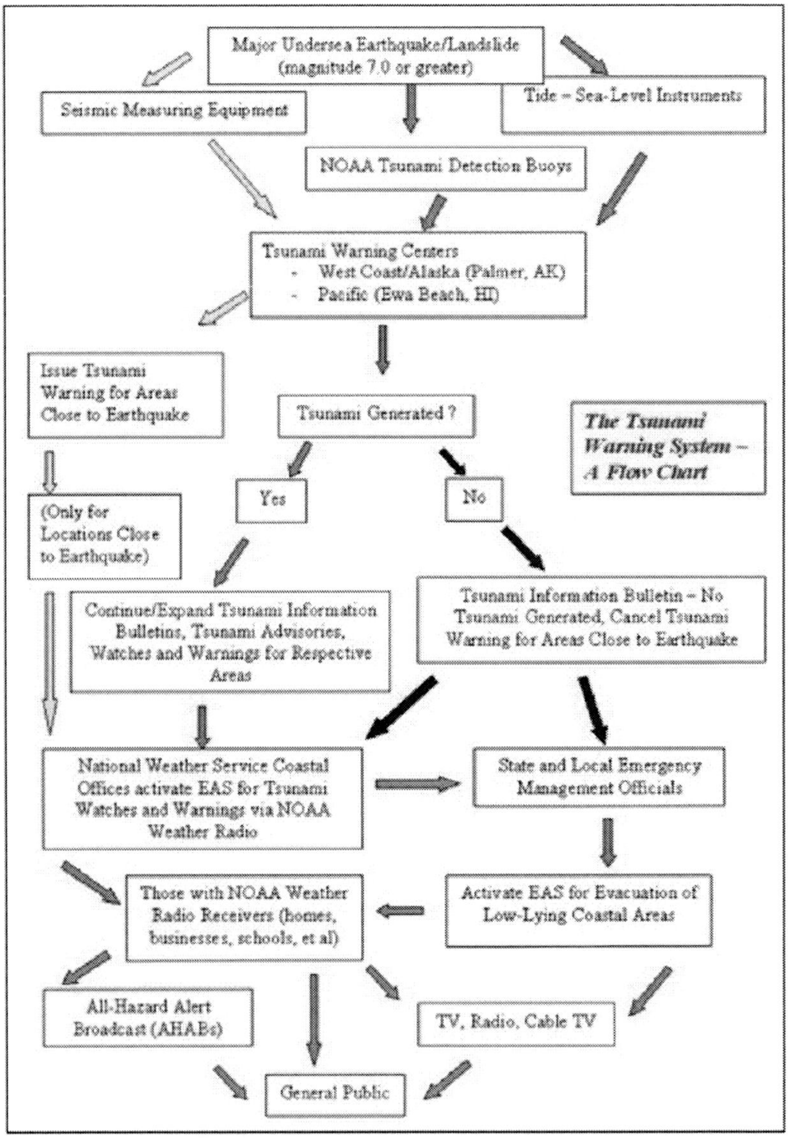

Source: NOAA, *How Does the Tsunami Warning System Work?* http://www.tsunami.noaa.gov/images/warning-system-smaller.jpg.

Figure 1. Flow Chart of the Tsunami Warning System.

The National Tsunami Hazards Mitigation Program

The National Tsunami Hazards Mitigation Program (NTHMP) assists states in emergency planning and in developing maps of potential coastal inundation for a tsunami of a given intensity. The NTHMP also operates tsunami disaster outreach and education programs through NOAA's *TsunamiReady* program. In 1992, NOAA launched the NTHMP to address the credibility of Pacific tsunami warnings and to reduce the number of "false alarms." The goal of NTHMP is to ensure adequate advance warning of tsunamis along all the U.S. coastal areas and appropriate community response to a tsunami.[6]

Detecting Tsunamis and Issuing Warnings

The tsunami warning centers monitor and evaluate data from seismic networks and determine if a tsunami is likely based on the location, magnitude, and depth of an earthquake.[7] If the center determines that a tsunami is likely, they transmit a warning message to NOAA's weather forecasting offices and state emergency management centers, as well as to other recipients. The centers monitor coastal water-level data, typically with tide-level gages, and data from NOAA's network of Deep-ocean Assessment and Reporting of Tsunamis (DART) detection buoys to confirm that a tsunami has been generated, and if not, to cancel any warnings.[8] A generalized decision tree network for the earthquake-detection-through-warning process is shown in Figure 1.

Warnings Triggered by the March 11, 2011, Tsunami

Initial warnings of an impending tsunami were first issued by the PTWC based on seismic information before the network of DART buoys and tide gages actually detected a wave generated by the earthquake.[9] According to NOAA, initial tsunami warnings are normally based only on seismic information to provide the earliest possible alert.[10] Because tsunamis travel more slowly than seismic waves, confirmation of a tsunami may take much longer than confirmation of an earthquake. That was the case for the March

11, 2011, tsunami. The DART network first detected the earthquake-triggered wave 27 minutes after the earthquake struck at 2:46 p.m. local time in Japan,[11] confirming that a tsunami had been generated and could lead to significant widespread inundation around the Pacific Ocean. Figure 2 shows results from a model depicting the tsunami wave propagation across the Pacific Ocean.

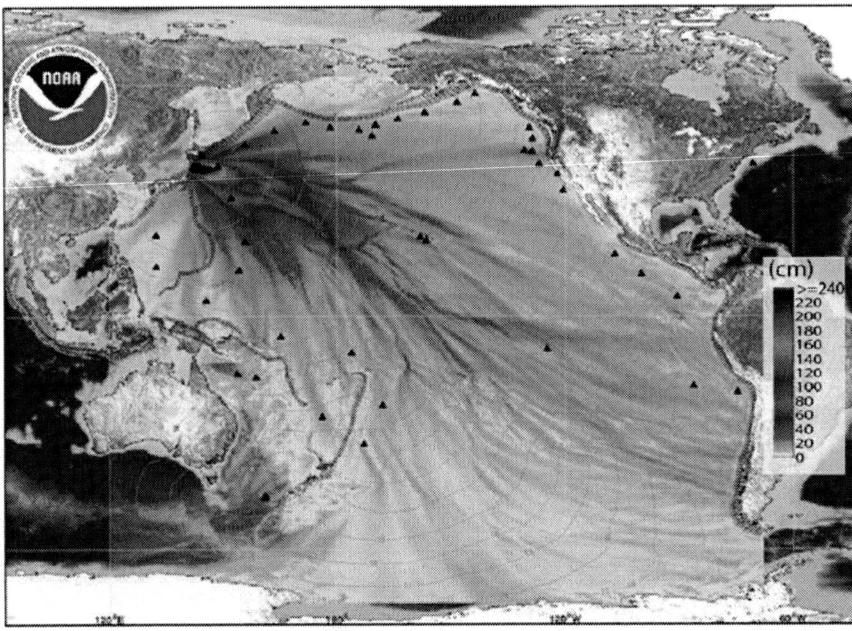

Source: NOAA Center for Tsunami Research, Pacific Marine Environmental Laboratory, http://nctr.pmel.noaa.gov/honshu20110311/.

Notes: Colors indicate the wave amplitude in centimeters (see scale bar on right side of figure); contour labels indicate the computed tsunami arrival times. Black triangles indicate location of Deep-ocean Assessment and Reporting of Tsunamis (DART) detection buoys.

Figure 2. Results from NOAA Model Depicting the March 11, 2011 Tsunami Propagating Across the Pacific Ocean.

The DART Buoy Network

NOAA first completed a six-buoy DART array in 2001 in the Pacific Ocean. Shortly after the 2004 Indian Ocean earthquake and tsunami that killed over 200,000 people, Congress passed H.R. 1674, the Tsunami Warning and

Education Act (P.L. 109-424), to enhance and modernize the existing Pacific Tsunami Warning System to increase coverage, reduce false alarms, and increase the accuracy of forecasts and warnings, among other purposes. In part, the 2004 tsunami provided the impetus to expand and upgrade the DART system and to improve the U.S. capability to detect and issue warnings for tsunamis generally. As a result, the array was expanded to a total of 39 DART buoys in March 2008.[12] (See Figure 3.)

Currently, 7 of the 39 buoys are not operational and in need of repair. Of the 7 buoys that are not working, 5 are deployed in the Pacific Ocean. Other countries also operate DART buoys in the Pacific (e.g., Australia and Russia), but if another U.S. DART buoy ceases to function less than 80% of the U.S. DART network would be operational. The Tsunami Warning and Education Act (P.L. 109-424) requires that NWS ensure that maintaining operations of tsunami detection equipment is the highest priority within the tsunami forecasting and warning program at NOAA. Further, P.L. 109-424 requires that the NOAA Administrator notify Congress[13] within 30 days of (1) impaired regional forecasting capabilities due to equipment or system failures; and (2) significant contractor failures or delays in completing work associated with the tsunami forecasting and warning system.[14]

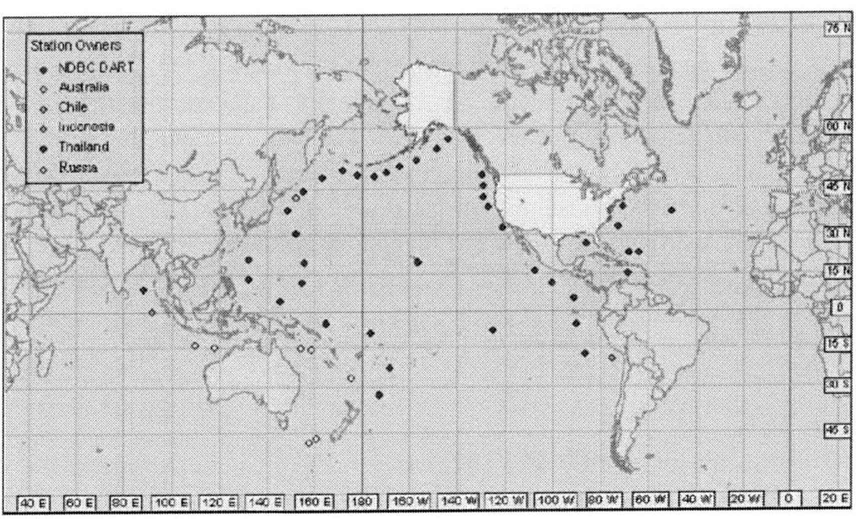

Source: NOAA National Data Buoy Center, http://www.ndbc.noaa.gov/dart.shtml

Notes: The United States owns and operates 39 of the DART Buoys.

Figure 3. Locations of DART Buoys.

Tsunami Warnings from the Japan Meteorological Agency

According to the International Tsunami Information Center, which operates under the International Oceanographic Commission (IOC)—part of the U.N.'s Educational, Scientific, and Cultural Organization (UNESCO)—the Japan Meteorological Agency (JMA) issued a major tsunami warning 3 minutes after the earthquake struck at 3:46 pm local time.[15] The first regional tsunami bulletins were issued by the North West Pacific Tsunami Advisory Centre (NWPTAC), operated by the JMA, about 9 minutes after the earthquake occurred.[16] The first tsunami wave reached the Japan coastline nearest to the epicenter about 15 minutes after the earthquake.[17]

The network of tsunami warning centers is coordinated under the umbrella of the IOC, through its Tsunami Programme, which falls under the auspices of UNESCO. According to the IOC, its role is coordinating the regional tsunami warning systems.[18] The IOC coordinates the Indian Ocean Tsunami Warning and Mitigation System (IOTWS), in addition to its role in the Pacific, per U.N. mandate after the 2004 Indian Ocean tsunami. The IOC also coordinates similar systems in the Caribbean (CARIBE-EWS) and the North-Eastern Atlantic and Mediterranean (NEAMTWS).

The IOC noted that for the March 11 earthquake the warning centers operated well and according to expectations: the seismic systems identified the location and magnitude of the earthquake within minutes and allowed for early warnings; the DART buoys confirmed the initial tsunami warnings and alerts; and the communication systems allowed for near-real time monitoring.[19] As a result, countries with Pacific Ocean coastlines received adequate warning in time to prepare for the oncoming tsunami waves. Northeast Japan, however, suffered the worst damage because it is so close to the epicenter, and the waves struck before people could evacuate to safety. In such instances, the ground shaking caused by the earthquake may be the only early indicator for people to act upon who live closest to the epicenter of an impending tsunami.[20]

Funding for the Tsunami Program

Funding for the NOAA tsunami program supports three main categories of activities: (1) *warning*, such as the activities of the tsunami warning centers and DART network; (2) *mitigation*, such as the activities of NTHMP; and (3) *research*, including activities conducted by the Pacific Marine Environmental Laboratory and the National Buoy Data Center.[21] In the NOAA budget, these activities are cross-cutting among different activities under the NWS line item.[22] GAO, which analyzed funding data for the three general categories, noted that total funding for all these activities ranged from $5 million to $10 million annually between FY1997 and FY2004, but increased after the 2004 Indian Ocean tsunami from approximately $27 million in FY2005 to $42 million in FY2009. According to GAO, the proportion of funding allocated to warning activities increased from about 40% of the total in FY2004 to approximately 70% of the funding in FY2009.[23] The proportion allocated to mitigation decreased from approximately 50% of the total in FY2004 to about 30% in FY2009, while the proportion for research remained steady between about 6% to 10%.

Funding for the NWS tsunami program for FY2010 was approximately $41 million, allocated as follows:

- $23 million—Strengthen U.S. Tsunami Warning Program;
- $13 million—Spectrum Auction funding;[24]
- $4 million—NWS/Local Warnings and Forecasts; and
- $1 million—Office of Oceanic and Atmospheric Research/Pacific Marine Environmental Laboratory.[25]

In 2010, the Government Accountability Office (GAO) found that NOAA had made progress since 2005 in expanding and strengthening its tsunami warning and mitigation capabilities, including the deployment of the 39 DART buoys. GAO also found that operating and maintaining the buoys has proved difficult and costly, consuming about 28% of the total NOAA Tsunami Warning Program budget in FY2009.[26] GAO noted that NOAA is exploring ways to reduce maintenance costs by improving buoy reliability.

According to NOAA, the current continuing resolution (P.L. 112-4) does not allow the NWS to allocate FY2011 funding to purchase ship time required to repair the seven DART buoys that are not working.[27] As noted above, the delay or failure in completing work associated with the tsunami forecasting and warning system by contractors should also trigger notification of Congress by the NOAA Administrator under P.L. 109-424.

End Notes

[1] A tsunami is a large ocean wave typically caused by a subsea earthquake or volcanic eruption that can cause extreme destruction when it strikes land.

[2] CNN U.S., *Tsunami Waves Reach Hawaii, Eye West Coast*, CNN Wire Staff, March 11, 2011, http://www.cnn.com/ 2011/US/03/11/tsunami/index.html?hpt=T1.

[3] Despite the tsunami warnings, some communities along the West Coast and in Hawaii suffered damages. For example, some boats and harbor facilities were damaged by the tsunami in Crescent City, CA, although most of the fishing fleet headed out to sea to avoid the waves before they reached the harbor, according to the Los Angeles Times. Crescent City has suffered tsunami damage in the past, particularly from the 1964 Good Friday earthquake that struck Alaska. See Maria L. La Ganga, "Crescent City Comes to Grips with Tsunami's Devastation," *Los Angeles Times*, March 13, 2011, http://www.latimes.com/news/local/la-me-japan-quake-crescent-city-20110313,0,5296998.story.

[4] U.S. Geological Survey, Earthquake Hazards Program, http://earthquake.usgs.gov/earthquakes/eqinthenews/2011/ usc0001xgp/#details.

[5] See NOAA, NWS, "How *TsunamiReady* Helps Communities and Counties at Risk," http://www.tsunamiready.noaa.gov/.

[6] NOAA FY2012 Blue Book, Chapter 5, National Weather Service, p. 691, http://www.corporateservices.noaa.gov/ nbo/fy12_presidents_budget/National_Weather_Service_FY12.pdf.

[7] Nearly all tsunamis are triggered by subsea earthquakes, although some may also be caused by underwater volcanic eruptions or landslides.

[8] U.S. Government Accountability Office, *U.S. Tsunami Preparedness: NOAA Has Expanded Its Tsunami Programs, but Improved Planning Could Enhance Effectiveness*, GAO-10-490, April 2010, p. 5.

[9] DART buoy 21418; telephone conversation with Laura Furgione, Deputy Director, National Weather Service, March 15, 2011.

[10] NWS, Pacific Tsunami Warning Center, *About PTWC Messages*, http://ptwc.weather.gov/ptwc/about_messages.php.

[11] Telephone conversation with Laura Furgione, March 15, 2011.

[12] According to NOAA, 33 of the DART buoys are deployed in the Pacific Ocean, and the rest are deployed in the Atlantic Ocean and Caribbean. NOAA National Data Buoy Center, Deep-ocean Assessment and Reporting of Tsunamis (DART) Description, http://www.ndbc.noaa.gov/dart/dart.shtml.

[13] Specifically, P.L. 109-424 requires the NOAA Administrator to notify the Committee on Commerce, Science, and Transportation in the Senate and the Committee on Science (now Science, Space, and Technology) in the House.

[14] The statute does not define what is considered impairment of the forecasting abilities, or what is a threshold for significant contractor failures or delays. However, the committee report accompanying the bill states that NWS is required to notify Congress when the tsunami forecasting capabilities are impaired for more than three months; U.S. Congress, House Science, *United States Tsunami Warning and Education Act*, report to accompany H.R. 1674, 109th Cong., 2nd sess., 2006, H.Rept. 109-698, p. 10. NWS uses an 80% operational threshold as its internal guideline; Telephone conversation with Laura Furgione, March 15, 2011.

[15] UNESCO, International Tsunami Information Center, http://itic.ioc-unesco.org/.

[16] World Meteorological Organization, March 11, 2011, http://www.wmo.int/pages/mediacentre/news/index_en.html.

[17] UNSESCO, Intergovernmental Oceanographic Commission, http://ioc-unesco.org/.

[18] UNESCO, International Tsunami Information Center, http://itic.ioc-unesco.org/.

[19] Ibid.

[20] Personal communication, Dr. Gregory van der Vink, seismologist, March 14, 2011.

[21] U.S. Government Accountability Office, *U.S. Tsunami Preparedness: NOAA Has Expanded Its Tsunami Programs, but Improved Planning Could Enhance Effectiveness*, GAO-10-490, p. 7.

[22] For example, the FY2010 enacted budget contains a line item: Strengthen U.S. Tsunami Warning Network—$23.264 million. However, research activities for tsunamis are included in the overall budget for the Pacific Marine Environmental Laboratory and for the National Buoy Data Center.

[23] U.S. Government Accountability Office, *U.S. Tsunami Preparedness: NOAA Has Expanded Its Tsunami Programs, but Improved Planning Could Enhance Effectiveness*, GAO-10-490, p. 8.

[24] Starting in FY2009, the tsunami program received funding from the proceeds of the Federal Communication Commission's auctioning of broadcast frequency spectrum. In FY2012, the program will be augmented by $12.7 million from auction proceeds, according to NOAA. Total funding received from auction proceeds will be approximately $50 million for the tsunami program at the end of FY2012, according to GAO.

[25] E-mail from Lara Hinderstein, NOAA Budget Outreach and Communications, March 11, 2011.

[26] U.S. Government Accountability Office, *U.S. Tsunami Preparedness: NOAA Has Expanded Its Tsunami Programs, but Improved Planning Could Enhance Effectiveness*, GAO-10-490, p. 21.

[27] Approximately $4 million would required, according to NOAA. Telephone conversation with Laura Furgione, March 15, 2011.

INDEX

D

E

F

G

P

R

S

T

U

V

W